Harmony between Architecture & Scenery

风景建筑

（卢森堡）瓦格纳·卢克 / 韦伯·乔治 编 常文心 译
Edited by Luc Wagner, Jörg Weber
Translated by Chang Wenxin

辽宁科学技术出版社

CONTENTS 目录

FOREWORD	4	前言	
INTRODUCTION	6	概述	
LANDSCAPE PAVILION		景观亭	
Definition	8	定义	
Functions	8	景观亭的功能	
Key points for designing a landscape pavilion	10	景观亭的设计要点	
Photovoltaic panels	11	光伏电板的使用	
PROJECTS		案例	
Phoenix Civic Space Shade Canopies	14	凤凰城市民活动区凉亭	
Deck over a Roman Site in Cartagena	18	卡塔赫纳罗马遗迹平台	
Vinaros Sea Pavilion	24	韦纳洛斯海滨凉亭	
Iron Bark Ridge Park Environment Centre	30	澳洲橡木山公园环境中心	
Ruth Lily Visitors Pavilion	34	卢斯·莉莉游客亭	
LANDSCAPE RESTAURANT/CAFE		景区餐饮建筑	
Landscape & Architecture	38	建筑与景观	
Serving sightseeing-type building	40	服务性建筑	
Tourist catering building	41	饮食业建筑	
PROJECTS		案例	
Riverside Restaurant	44	天门山"山之港"临江餐厅	
HOTO FUDO	50	HOTO 餐厅	
The Terrace View Café	56	露台咖啡厅	
Les Grandes Tables de L'île	62	岛上大餐桌	
Pavilion Madeleine	68	玛德琳餐厅	
VISITOR CENTRE		游客中心	
Guiding criteria	72	指导标准	
Visitor centre proposals	74	游客中心的建设	
Site design	83	规划设计	
Environmental considerations	90	环境因素	
Cultural considerations	93	文化因素	
Architectural design	94	建筑的设计	

Relations between architecture and the surrounding landscape	108	建筑和周围景观设计的关系
Visitor centre and tourist safety	109	游客中心和旅游安全
Sustainable design considerations	112	可持续设计因素
Cradle-to-grave analysis	113	长期目标的分析
Visitors and poisonous plants	115	游客与常见有毒植物
PROJECTS		**案例**
Pavilions in Lake Yangcheng Park, Kunshan	118	昆山阳澄湖公园景观建筑
GATEWAY to the McDowell Sonoran Preserve	122	麦克道尔·索诺兰保护区入口
Botanical Gardens Chenshan	126	辰山植物园
UNESCO World Natural Heritage, Messel Pit Visitor Information Centre	134	世界自然遗产——麦塞尔化石坑游客信息中心
Tibet Linzhi Namchabawa Visitor Centre	140	西藏林芝南迦巴瓦接待站
Queens Theatre in the Park	146	女王剧场
Visitors Pavilion at Ramat Hanadiv	152	莱姆特·哈那迪夫游客亭
Urban Redevelopment of the Plaza del Milenio	160	千禧广场城市改造
Red Rock Canyon Visitor Centre	166	红岩谷游客中心
Welcome Centre of Alphaville	172	阿尔法城迎宾中心
Mirror House at Copenhagen Central Park	178	哥本哈根中央公园镜子屋
VanDusen Botanical Garden Visitor Centre	182	范杜森植物园游客中心
Thomas Jefferson Visitor Centre and Smith History Centre at Monticello	188	托马斯·杰弗逊游客中心和史密斯历史中心
Arche Nebra	194	内布拉方舟
Bai Sha Wan Beach and Visitor Centre	198	白沙湾海水浴场旅客服务中心
Hanil Cement Visitors Centre and Guesthouse	204	韩一水泥游客中心和宾馆
Vanke Triple V Gallery	208	万科 3V 画廊
Snaefellsstofa, Visitor Centre	214	斯纳菲尔斯托法游客中心
New Visitor Centre of the Pannonhalma Arch-abbey	218	帕农哈尔玛大修道院游客中心
Cabrera Islands National Park Visitors Centre	224	卡布雷拉岛国家公园游客中心
PAV – Park of Living Art	232	生活艺术公园
Pavilion in Grebovka	236	戈罗波维卡亭
2010 Taipei International Flora Exposition	242	2010 台北国际花卉博览会
Sun Moon Lake Administration Office of Tourism Bureau	248	日月潭风景区管理处
Edithvale Seaford Wetlands Discovery Centre	254	艾迪斯维尔西弗德湿地发现中心
Dashawan Beach Facility	260	大沙湾海滨浴场
Architecture of Discovery Green	264	发现绿色公园景观建筑

FOREWORD

Partner WW+
Luc Wagner, urban planner and **Jörg Weber**, architect
WW+
Head office Luxembourg
21, rue Jean Jaurès
L – 4152 Esch-sur-Alzette
T +352 26 17 76
Office Germany
Südallee 41 b
D – 54290 Trier
T +49 651 99 98 90 00
E info@wwplus.eu
W www.wwplus.eu

Dialogue between Architecture and Landscape

Architecture and Landscape – A Human Need
Planning per se is the imaginary anticipation of a future desired state, strongly affected by the different regional influences and also by the time-period in which the planning is 'taking place'. This is the particular subjective intention of the architect or landscape architect.

Architecture on the one hand, as a confrontation between man and his built structures and landscape on the other hand, in terms of a cultural landscape, both were established as a result of a creative design process. This is triggered by the needs of human beings for protection from environmental influences, as well as their necessity of supply through e.g. agriculture or forestry, fresh water and other requirements.

The Dialogue – A Human Skill for Development
Based on David Bohm's interpretation of a group-dialogue, seeing a dialogue as more than communication between people but also as a transformation of them, William Isaacs recorded the following elementary dialogue skills: Listening, Respect, Suspension, Articulation. [1] Transferring Isaacs' qualities for a dialogue from the issue of human beings in general onto the creative process of planning and its built results, this would mean:
Listening in terms of recognising the 'genius loci', the 'spirit of place' with its typical features, patterns or structures, reading the history and atmosphere of a place in its cultural uniqueness and temporal dimension;
Respect as the conscious use of the existing situation, respect for the recognised 'genius loci' by sustainable and ecological planning, while maintaining elements being worth conserving. It is also an acknowledgement of the local planning requirements, which are developed out of the currently determined design guidelines of culture and time;
Suspension in sense of loosening or detaching as a detection and monitoring of own thoughts, emotions and opinions;
Articulation standing for the planning procedure as well as the planned objects as a result of the finding process in the style of each creative individual.

The Dialogue – Enriching Architecture and Landscape
Having a dialogue between architecture and landscape, both parts will be mutually enriched, even if one or the other may be predominant. Establishing a dialogue means augmenting the uniqueness of a built environment, existing only once and having an

前言

风景建筑——建筑与景观的对话

建筑与景观——人类的需求
规划本身是对未来理想状态的期望,受到地域和规划时间的强烈影响。这是建筑师或景观建筑师特有的主观愿望。

建筑作为人类与其建成结构之间的碰撞,与文化景观一样,两者都是创意性设计过程的结果。这是由人类对于免受环境影响的需求以及他们对从农业、林业、淡水等方面获取必需品的要求而引发的。

对话——人类开发的技能
威廉·艾萨克记录了以下基本的对话技能:1.听,2.尊重,3.停顿,4.发音。[1]将艾萨克总结的对话特点转化为全体人类与创意规划流程及其建成结果之间的对话,即意味着:
"听"意味着辨识"地方特色"、地方精神(如典型特征、模式或结构)、阅读历史和地方氛围(包含文化特色和时间维度);

"尊重"是有意识地利用现有条件,通过可持续和生态规划尊重已经认知的"地方特色",同时保留值得保护的元素。它也是一种对当地规划需求的认证,通过已经决定的文化和时间设计原则来进行开发;
"停顿"意味着放松或脱离,以便发现并监督我们自己的思想、情绪和意见;
"发音"代表着规划流程和规划目标,找寻各个创造个体的风格。

对话——丰富建筑与景观
在建筑与景观之间开展对话,无论是处在支配地位,还是主要地位,二者都会得到丰富。建立对话意味着扩大环境的独特性,挖掘它成为和谐整体的潜力。

建筑和景观的融合可以通过普通的材料、质感、色彩或造

individual character with its potential opportunities being a harmonious whole.

A dialogue can be generated by using a common materiality, texture, colour or a common formal language. The fusion of inside and outside as well as surprising views and contrasting landmarks can also be a way of establishing a dialogue. The time component adds a further dimension to the interactive process between architecture and landscape.

The Dialogue – Past and Present
After medieval Europe the art of gardening, as a forerunner of landscape architecture, was strongly linked to the design of representative buildings. This link faded in modern times, only to be discovered today in recent years although in a less formal way. Contemporary architectural language places the importance of creating unities on interpreting past and valuable principles rather than copying old models.

Peeking into China's past we find a basic principle of an existing dialogue: A traditional Chinese home, a 'Siheyuan', often had a courtyard with a tree in its centre. The size of a grown single tree is just appropriate to cover the entire courtyard and give a shelter. Even in a small scale concept, both architecture and landscape can function as a unit, in order to create places appropriated for human beings.[2]

Looking at China's present shows us that urban voids are occupied in everyday life in full measure, even if their function is not clearly defined: The Chinese community is capable of adopting public space; they would never allow any land to remain unoccupied. This is partly due to the high population density of cities in the country. In the old cities of a megalopolis like Beijing for example, since the housing is fairly limited, residents consider public space as an extension of home, inhabitants put their chairs and tables in the street, under the trees, take a tea or play a game of chess.[2]

The Dialogue – Our Future
Our goal as architects and urban planners is to work interdisciplinary in a team of landscape architects and representatives of other disciplines such as energy consultants, artists, traffic planners, etc. to develop a coherent overall strategy to allow a high identification of the people with their working, living and leisure sites in order to increase the quality of life.

[1] Dialogue Project 1992-1994 MIT- Massachusetts Institute of Technology, USA
[2] XIAO Jue, Chinese architect, freelance in 2011 at WW+, Luxembourg

型语言产生。时间元素进一步促进了建筑与景观之间的对话。

对话——过去和现在
中世纪以后，作为景观建筑的先驱，欧洲的园艺艺术与代表性建筑的设计紧密联系在一起。这种联系在现代逐渐消失，近几年仅以非正式的形式呈现：现代建筑语言将重点放在诠释过去的统一和创造价值原则上，而不是全盘复制旧模型。

审视中国的过去，我们发现了一个基本原则："传统的中式住宅——四合院的中央通常设有栽种树木的庭院。高大的树木正好能覆盖整个庭院，提供了避难所。尽管这一设计规模很小，建筑和景观能够作为一个整体为人类打造合适的空间。"[2]

审视中国的现在我们发现城市留白已经被日常生活全面占领，尽管它们的功能尚不明确："中国社会善于采纳公共空间；他们绝不会让任何土地空闲。这在一定程度上是由城市的人口密度所决定的。例如，在北京这样人口稠密的古老城市中，由于住宅十分有限，居民将公共空间视作家庭的延伸，他们将桌椅摆在街上、在树下饮茶或下棋。"[2]

对话——我们的未来
建筑师和城市规划师的目标是与景观建筑师和其他学科的代表（例如能源顾问、艺术家、交通规划师等）进行跨学科的合作，共同打造一个整体策略，让人们的工作、生活和休闲场所具有高度的辨识性，以提高生活的质量。

[1] 对话项目1992-1994，美国麻省理工学院
[2] 肖珏，中国建筑师，2011年WW+卢森堡总部自由撰稿人

INTRODUCTION

风景建筑概论

Landscape building possesses two basic features, one is the matching of architecture style with the beautiful scenery, which not only requires the harmony with the surrounding natural environment, but also adds glamour to the environment; the other is, its site should satisfy the functional requirements of both and enjoying the view.

Relations between the surrounding natural scenery (environment)

Landscape building transforms the natural scenery according to its existing typography, or creating the scene artificially, combine the plants and layout of the building, which embodies the intimate relations between pavilion and nature.

Relations between local cultures

Most of the scenery spot is in the local historic cultural background. Thus, landscape building should consider the relations with the local cultures. The traditional historic scenes always possess rich culture heritage.

Colour of the architecture

The architecture colour is the key embodiment of the landscape building style. Meantime, the colour can be understood from two aspects. One refers to the exterior colour of the architecture; the other is the entire reflection of the landscape vision formed by the building skirts. The architecture in the scene should be avoided the vision pollution, while colour is the perfect embodiment of architecture style.

风景建筑要具备两个基本特征，一是建筑造型与优美的风景相匹配，不仅要求与周围的自然环境相和谐，更要为环境增色；二是它的选址要同时满足点景与观景的功能要求。

与周围自然风景（环境）的关系

风景建筑在一定的范围内，利用并改造了天然山水地貌或人为开辟山水地貌，结合植物的栽植和建筑的布局，体现建筑和自然水乳交融的关系。

与地域文化的关系

大多数风景区都处在地区性的历史文化背景之中，因此，风景建筑要考虑与地域文化的关系。传统历史环境中的风景，一般都有丰富的文化积淀。

建筑的色彩

建筑色彩是景区建筑风格的重要体现。同时，色彩可以从两方面理解，一方面是指建筑物的外装饰颜色，另一方面是指建筑群体结合成的景观效果在视觉中的整体反映。景区内的建筑，需要防止的是造成视觉上的污染，而色彩是建筑物格调完美体现的点睛之笔。

Landscape Pavilion

景观亭

1. Definition

Landscape pavilion refers to a small architecture which is exquisite and detailed, providing a rest space for visitors appreciating views. All around of the pavilion are open. They are comparatively small and intensive, possessing comparatively independent architectural images, light and flexible. They are often combined with the surrounded mountain, water and green space, enriching and enlivening the whole space, which becoming one of the sights of landscape. Landscape pavilion not only has rich cultural connotation, but also highlights the local landscape features. The reasonable design and arrangement of landscape pavilion contribute to fabulous artistic effect of the landscape space.

2. Functions

2.1 Rest

To provide the visitors with a space for resting and sheltering while enjoying the views.

2.2 Enjoy the scenery

As a viewpoint for sightseeing and enjoying the view, the choice of the location of a pavilion aims to satisfy the viewing distance and perspective. It can be located at the top of a mountain, on the roadside, in the water, or at the end of a bridge.

• Hill pavilion is suitable to be located at the terrain which is good for looking far into the distance, for instance, the top of the hill, the ridge, covering a wide range and many directions. The hill and the pavilion should be in proportion to the size and height, which should make good use of the geography characteristics, adopting various architectural structure forms to enrich the hill landscape, thus, the hill is more vivid, meantime, the climber may stop to enjoy the comfortable environment for a rest.

• Building the pavilion can not only enjoy the water scenery, but also enrich the waterscape. The size of the water pavilion depends on the amount of water. To highlight the different landscape effect, it is preferable to be built the pavilion low to the water, while on the broad open water, the pavilion is better to build on the high platform or higher rocky ledge to enjoy the distance mountain and the near water. On the islands, both the heart of lake and the rock near the shore are the location of building the

1. 定义

景观亭，是指景观绿地中精巧细致可供游人休憩及赏景的小型建筑物，其四周开敞，造型相对小且集中，有相对独立的建筑形象，轻巧、灵活，并常与周围的山、水、绿地相结合，充实活跃整个空间，成为景观中的一景。它既具有丰厚的文化内涵，又能凸显当地景观特色。

景观亭的合理设计与安排，可构成景观空间中美好的景观艺术效果。

2. 景观亭的功能

2.1 休息

满足人们在游赏活动的过程中驻足休息、纳凉避雨，是游人休息之处。

2.2 赏景

作为景观中凭眺、观赏景色的赏景点，亭的位置选择应主要满足观赏距离和观赏角度的要求。它既可分布在山巅、路旁，也可安置在水中、桥头。

• 山亭宜建于适合远眺的地形之上，如山巅、山脊，其眺望的范围宽广、方向多。山与亭在体量及高度上应比例协调，充分利用地形环境特点，采用各种建筑结构形式，以增加山体景观，使山色更有生气，同时为登山中的休憩提供一个可停下观赏的舒适环境。

• 临水建亭既可观赏水上景色，又可丰富水景效果。水亭的体量主要取决于水的大小。为突出不同的景观效果，一般在小水面建亭宜低临水面，而在碧波坦荡的大水面上，亭宜建在临水高台或较高的石矶上，以观远山近水。在小岛，湖心台基上、岸边石矶上都是临水建亭之处。在桥上建亭，更能使水面景色锦上添花，并增加水面空间层次，桥上建亭还需与桥身及周围建筑风格相统一，为整个景观

pavilion near the water. Building the pavilion on the bridge even adds brilliance to the beautiful waterscape, which can also add the space layer of the water. Building on the bridge should also be in accordance with the surrounded architecture style to add more glamour to the whole landscape.

· Locating the pavilion on the ground is usually the key point of the road, for instance, the square, the crossroad and the avenue of the sideways. Some natural scenery spot place a pavilion in the centre or on the sideway as a landmark and destination. The style, material and colour of the pavilion should be taken into consideration with the specific environment. Building the pavilion on the ground is often combined with the road, and its location should avoid the main stream, not on the traffic trunk road. The ideal location is sideways, crossings or branch roads. Some of them require to be leveled higher. To avoid the dull and pinched situation, the pavilion should be built in accordance with the surrounded landscape, with the decoration of stones and flowers. The style of the pavilion itself should also be new and attractive.

Located at the the Plaza de la Encarnacíon in Seville, 'Metropol Parasol' is designed by J. MAYER H. architects. Its highly developed infrastructure helps to activate the square. The main structure is a tremendous 'Parasols' formed by a set of giant honeycombed wood structure. The parasols grow out of the archaeological excavation site into a contemporary landmark, the organic curves and ballooning forms of these 'parasols' contrast with the more mundane rectangles surrounding this former parking lot in Seville, defining a unique relationship between the historical and the contemporary city. (Figure 1)

锦上添花。

· 平地设亭多位于道路中的重点部位，如广场、道路的交叉口、路侧林荫之间。有的自然风景区在进入主要景区之前，在路边或路中筑亭，作为一种标志和点缀。亭子的造型、材料、色彩要与所在的具体环境统一起来考虑。平地建亭常与道路相结合，其位置应退出人流路线，不要在通车干道上，选择在路侧、路口或小岔路中，有时需适当提高亭的基址。平地建亭，为避免平淡、闭塞，要结合周围环境造成一定景观效果，借助于布石、栽花加以点缀。亭子本身的造型也应新颖诱人。

位于塞维利亚卡尔纳西恩广场的"都市阳伞"项目由J. MAYER H. 设计事务所设计，其发达的基础结构活跃了广场，工程主体部分是一组巨大的蜂窝状木质结构，它形成的巨大"阳伞"，从考古发掘场地伸展而出，它的有机弧线和膨胀造型与周边平凡的直线型结构形成了鲜明对比，在历史和现代城市之间建立了独特的联系。（如图1）

Figure 1. 'Metropol Parasol' is designed by J. MAYER H. architects.
图1 "都市阳伞"项目由J. MAYER H. 设计事务所设计

Landscape Pavilion

2.3 The pavilion is in harmony with its surounding landscape
The location, style, size and colour of the pavilion are combined with the surrounded landscape accordingly, which draw the outside bigger space into it. Through the correct estimation of the present situation, we should make best use of the advantages and bypass the disadvantages to achieve the best environmental landscape composition effect.

2.4 Specific functions
Some pavilions have specific functions, like Memorial Pavilion, ticket booth, food booth and so on.

3. Key points for designing a landscape pavilion

3.1 A good location should be firstly taken into consideration. Make good use of its advantages like its small size and less likeliness to be influenced by location, topography and so on.

3.2 The choice of the size and pattern of the pavilion mainly depends on the size and nature of the surrounded environment accordingly. The type of modern landscape pavilion is livelier and the forms are much more than the old ones, including flat type pavilion, umbrella pavilion and mushroom pavilion. It prefers the organic combination with the surrounded environment. The design itself not only has good proportion and exquisite look, but also coordinates and unifies the architecture with its surrounded environment in size and type. And the design should be appropriate and natural, neither inadequate nor too much. Oversize may destroy the wholeness of the landscape.

3.3 The materials of the pavilion prefer to draw on local resources, which is good for processing and closer to natural design.

The type of landscape pavilion depends on the materials to some extent. Owing to the differences of the materials, the pavilion carries unique features with distinctive materials, at the same time; they are also limited by the materials they used. Earlier landscape pavilion used to choose wood, bamboo, stone, thatch, steel reinforced concrete as the main resources. Recent years, grass, metal, plastic resin, shell inflatable

2.3 点景
景观亭的位置、造型、体量、色彩等因地制宜的与周围景观相结合，把外界大空间的景象吸收到这个空间中。通过对现有环境特征和景观做出正确的分析评论，扬长避短，对环境景观构成中的不利因素进行外部形态的加工和再造，以获得最佳的环境景观构图效果。

2.4 专用
具有特定功能的景观亭，如纪念亭、售票亭、售货亭等。

3. 景观亭的设计要点

3.1 首先应选择好位置，发挥亭的占地较小、受方位、地形等影响小的特点，按照总体规划意图选址。

3.2 景观亭的体量与造型的选择，主要取决于它所处的周围环境的大小、性质等，因地制宜而定。现代景观亭在造型上更为活泼自由，形式更为多样，其中包括各种平顶式亭、伞亭、蘑菇亭等；在布局上更多地考虑与周围环境的有机结合。不仅应使其本身比例良好、造型美观，而且还应使建筑物在体量、风格等方面都能与它所在的环境相协调和统一，在处理上，要恰当、要自然，不要不及，更不要太过。超过了环境所允许的建筑体量也会损害景观的整体性。

3.3 亭子的材料应力求使用地方性材料，就地取材，不但加工便利而且也近于自然设计实例。

景观亭的造型从某种程度上取决于所选用的材料。由于各种材料性能的差异，不同材料建造的凉亭，就各自带有非常显著的不同特色，而同时，也必然受到所用材料特性的限制。早期景观亭的材料多以木材、竹材、石材、茅草、钢筋混凝土为主，近年，玻璃、金属、塑料树脂、薄壳充

Figure 2. 'Orquideorama' is designed by Plan B Arquitectos and JPRCR Arquitectos.
图2 "Orquideorama" 项目由建筑师Plan B建筑师事务所、JPRCR建筑师事务所设计

soft materials together with other new materials and new technology are brought into this architecture, which makes the landscape pavilion modern and fashionable.

Orquideorama is one of the ecological pavilions among Medellin botanical gardens in Columbia, which is designed by Plan B Arquitectos and JPRCR Arquitectos. Steel and local wood are the main raw materials. Steely structure is surrounded by the wood, which combine the architecture and organisms efficiently. From the micro perspective, geometry design and material structure make the building itself the nature of life. From the macro perspective, the whole building has a strong visual effect for each monomer connects with each other through cellular geometric shapes, systematically repeat, continuous extension, and integrate smoothly into thick plants. (Figure 2)

4. Photovoltaic panels

The creation of a new public civic space in Downtown Phoenix offered an exciting opportunity to give the city a new civic identity, but would require extensive shading to make the site hospitable. The concept by the office Architekton for the shade canopies was derived from the undulating topography of the park: understanding the ground as a single, warped surface. The intent was to create a parallel, overhead plane which would unify one's experience of the site. By placing the structural frame above the shading elements, a plane of metal conduit is able to read as a smooth, continuous underbelly, legible from beneath.

Photovoltaic panels placed on top of the structure provide further shade, and enhance the layered quality of the shadows cast below, which form an intricate and continuously changing pattern throughout the day.

'Photovoltaic' is a marriage of two words: 'photo', from Greek roots, meaning light, and 'voltaic', from 'volt', which is the unit used to measure electric potential at a given point. Photovoltaics (PV) is a solid-state technology that converts solar radiation directly into electrical power. Solar photovoltaic systems need only daylight so can work in most locations. Because the output from photovoltaic cells changes based upon the amount of daylight available, they will always be more efficient when placed in unshaded areas, facing at least partly South–facing; Output is dependent on the strength of the sunlight, not on the outside air temperature.

气软材料等新材料、新技术也被人们引入到这种建筑上，使得景观亭有了现代的时尚感觉。

Orquideorama是哥伦比亚麦德林市植物园里的一个生态建筑亭，由建筑师Plan B建筑师事务所、JPRCR建筑师事务所设计。它主要以钢材和当地的木材为原料，钢材为骨架，外面是木材包边。它将建筑和有机生命体有效的结合起来。从微观来看，自定义的几何图案以及材料的组织结构都让建筑本身具有一种生活的性质；从宏观上看，整个建筑有一种很强的视觉效应，每一个单体采用了蜂窝的几何形态而连在一起，有系统地重复，不断的延伸开来，跟茂密的植物很好的融合在一起。（如图2）

4. 光伏电板的使用

凤凰城市民活动区凉亭使用了新能源，它的建造给这个城市提供了一个新的身份，成为受游客欢迎的场所。建筑事务所活动区凉亭的设计理念来源于公园起伏的地形：把该地理解成一个独立变形的表面。目的就是建造一个平行，架空的水平面。通过凉亭上面的结构框架，人们能够从下面欣赏到金属导体光滑连续的表面。

顶端安置的光伏电板提供了更多的阴凉，并且增强了投射到下面的阴影处的层次质量，从而一天之中形成了错综复杂而又持续的变化类型。

"光电"一词来源于另外两个单词，希腊词根"照片"和来源于"volt"的"voltaic（电流的）"。光电（PV）是使用电晶体的一种科技，将太阳辐射直接转化成电力。太阳光电系统只需要日光，所有大部分地区都可以使用，因为光电细胞的输出变化取决于日照的多少。它们在阳光下的工作效率通常会比较高，至少也要部分朝南；输出量取决于日照的强度，与外面的空气温度无关。

Landscape Pavilion

The most common semi conductor material used in photovoltaic cells is silicon, an element most commonly found in sand. There is no limitation to its availability as a raw material; silicon is the second most abundant material in the earth's mass. Moreover a photovoltaic system does not need bright sunlight in order to operate. It can also generate electricity on cloudy days.

PV cells are generally made either from crystalline silicon, sliced from ingots or castings, from grown ribbons or thin film, deposited in thin layers on a low-cost backing.

Solar PV is an almost maintenance-free renewable energy source, because it is based on solid-state semiconductor technology. The inverter is designed to operate automatically to optimise the output of the system.

The most important feature of solar PV systems is that there are no emissions of carbon dioxide - the main gas responsible for global climate change - during their operation. Although indirect emissions of CO_2 occur at other stages of the lifecycle, these are significantly lower than the avoided emissions. PV does not involve any other polluting emissions or the type of environmental safety concerns associated with conventional generation technologies. There is no pollution in the form of exhaust fumes or noise.

Recycling of PV modules is possible and raw materials can be reused. As a result, the energy input associated with PV will be further reduced. (Figure 3/4/5/6)

硅是光伏电池中最常用到的半导体材料。这种元素通常在沙子中提取出来。硅作为原材料没有任何限制；而且硅也是世界上第二大丰富资源。更有利的是，光电系统不需要在明亮的日照下才能运行，它在阴天的时候也能够产生电力。

光电池通常提取于晶体硅，或是切割于铸铁或铸件，成本低廉。

太阳能光电可谓是无需维修、可再生的能量来源，因为它是建立在电晶体的半导体技术。变频器可以自动提升系统的输出量。

太阳能光电系统最重要的特征就是运行过程中没有二氧化碳的释放，它是造成温室效应的主要气体。尽管在其他程序中会间接排放一些二氧化碳，但是它们的数量要小得多。与传统的发电科技相比，光电所产生的污染物比较少，更为环保，没有废气或是噪音产生。

光电模块可以再利用，原材料也可以反复利用。因此，以光电形式的能量输入将会得到更加广泛的应用。(如图3-图6)

Figure 3(Left). Civic Space in Downtown Phoenix, see page 14
Photovoltaic panels placed on top of the structure
Figure 4(Middle) Figur 5/6(Below). Photovoltaic panels' details
图3（左）：凤凰城市民活动区凉亭的顶端安置的光伏电板，参见14页
图4（中），图5、图6（下）：光伏电板应用细部

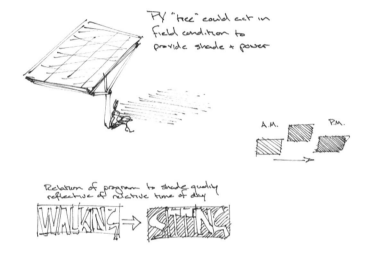

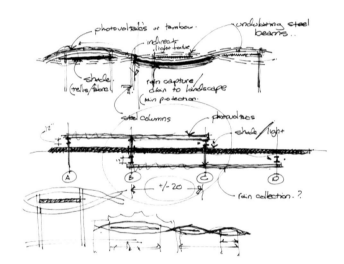

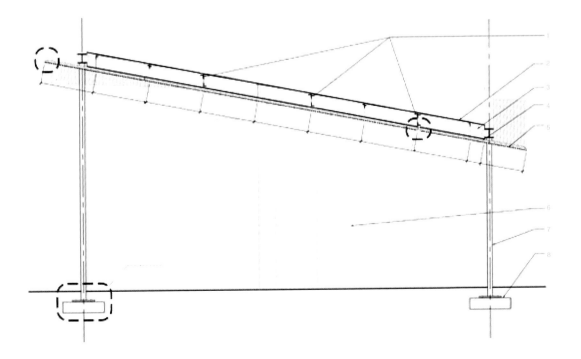

1. Lateral bracing
2. Photovoltaic panel
3. Steel beem
4. Unistrut part p4004 (TYP.) galvanized finish
5. 1*EMT
6. Light sticks (beyond)
7. Steel column
8. Footing

1.横向支架
2.光电能转换板
3.钢梁
4.镀锌Unistrut型钢
5.管道支撑系统
6.发光棒
7.钢柱
8.支撑点

Phoenix Civic Space Shade Canopies

凤凰城市民活动区凉亭

Completion Date: 2009 **Location:** Phoenix, Arizona, USA **Site Area:** 11,331sqm **Construction Area:** 1,328 sqm **Designer:** Architekton (Tempe, Arizona, USA)/EDAW | AECOM (Phoenix, Arizona, USA) **Photographer:** Winquist Photography/Eric Vollmer—Architekton/Michael T. Masengarb—Architekton Architekton/Daniel Watts—Next World Media

竣工时间：2009年 项目地点：美国，亚利桑那州，凤凰城 占地面积：11,331平方米 建筑面积：1,328平方米 设计师：阿奇泰克通公司（美国，亚利桑那州，坦佩），EDAW | AECOM（美国，亚利桑那州，凤凰城）摄影师：文奇斯特摄影；埃里克·沃尔默尔——阿奇泰克通公司；迈克尔·T·马森加伯——阿奇泰克通公司；丹尼尔·瓦特——异世界媒体公司

The Civic Space acts as Phoenix's 'central park', a gathering place for the community and a hub of Arizona State University's downtown campus. In a city bathed in sunshine over 330 days every year, it was imperative to provide a series of shade canopies that define outdoor rooms and give the project a singular cohesive identity. Playing off the arching planters that weave across the site, these canopies were envisioned as undulating surfaces that appear to flow seamlessly overhead.

Critical to the expression of thin floating planes was achieving a compact section. With a tightly compressed structural system, a custom unistrut profile allowed a minimal knife edge to cantilever well past the steel beams and hide the depth of the primary structural elements. Using vertical cantilevers to absorb lateral loads allowed the columns to obtain an absolutely minimal dimension.

In order to convey the impression of fluidity, a tight pattern of electrical conduit was suspended below the superstructure, creating a filigreed shade that would be suitable during the winter as well as the summer. Parallel girders set at opposing angles resulted in an incremental rotation of the structural plane, creating an overall hyperbolic shape out of purely linear components. A flexible connection detail allowed for easy installation of the 20,000 hangars and 16 miles of green painted rigid conduit that form the billowing under-surface of the shade structures.

Above the canopies, cross bracing serves as a substructure for a photovoltaic array, allowing the panels to gradually roll with the warp of the structure. Each panel is mounted at a unique angle, giving the impression of scales tilting towards the southern sun. This layer of photovoltaics not only serves to provide power for the park's lighting and create additional shade, but acts as a visible commitment to sustainability by the city.

这个市民活动区相当于凤凰城的"中央公园"，是社区集会的地方，也是亚利桑那州大学市中心校区的中心。在这座一年330天都沐浴在阳光之中的城市，凉亭是必不可少的设施，既可以界定露天空间，又能赋予项目凝聚力。凉亭与周边的弧形花池相匹配，穹顶呈流畅的波浪形。
薄薄的浮动面板形成了紧凑的截面。扁平的压缩结构系统让定制的单结构轮廓以最小的刃口边缘来支撑钢柱，同时也隐藏了基本结构元素。垂直悬臂结构吸收了横向荷载，将立柱的尺寸减到最小。
为了体现流畅性，紧密的电缆图案悬在上层结构之下，创造了冬夏皆宜的阴凉空间。对角而设的平行梁增加了结构面的旋转，以纯粹的线条元素形成了双曲线造型。柔性连接的细节设计让20,000根悬吊结构和16英里的绿色刚性导管（形成波浪般的遮阳结构下表面）的安装变得简便。
凉亭上方的交叉撑系是光电伏阵列的下层结构，让电池板随着结构的变形而逐步滚动。每块电池板都有其独特的角度，朝向南面的太阳倾斜。光电池层不仅为公园的照明提供能源、提供额外的阴凉，还体现了城市对可持续设计的追求和承诺。

Awards:
2010 AIA Arizona Citation Award
2009 McGraw-Hill Construction Best of the Best National Urban Design Award
2009 Southwest Contractor Best Urban Design Award
2009 Valley Forward Environmental Excellence Crescordia Award

获奖情况2010美国建筑师协会亚利桑那分会提名奖
2009麦格劳希尔集团最佳国家城市设计奖
2009西南承包商最佳城市设计奖
2009山谷先锋环境杰出奖

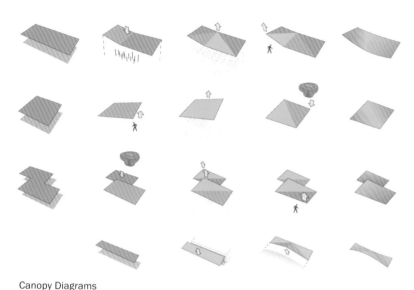

Canopy Diagrams
凉亭图

1. Night view of Shade Canopies
2. At nightfall, the pavilions in front of the buildings provide a gathering place.
3. Critical to the expression of thin floating planes was achieving a compact section.

1. 凉亭夜景
2. 傍晚大楼前的亭提供了聚会的场所
3. 薄板的设计呈现出紧凑的切面效果

4. The pavilions prevent people from the scorching sun.
5. The ground paving under the pavilion
6. The pavilion surrounded by flowers and trees
7. A bird's eye view of the whole structure and the surroundings

4. 在亭下休憩的人们免除了烈日的困扰
5. 亭下的地面铺装
6. 鲜花、绿树环绕的景观亭
7. 俯视整个建筑结构和周围景色

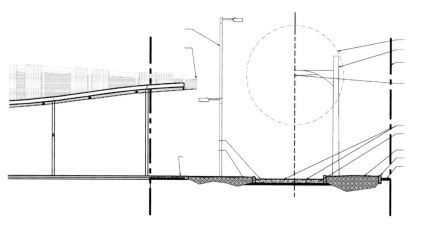

1. Photovoltaic panels
2. Support substructure
3. Primary steel superstructure
4. Unistrut sublayer
5. Conduit scrim

1. 太阳能光电板
2. 辅助结构
3. 基础上部钢结构
4. 焊件次级层
5. 导管平纹麻布

Site Plan
1. Post office canopy
2. Stage canopy
3. 424 canopy
4. Light stick canopy
5. Entry canopy

总平面图
1. 邮局凉亭
2. 舞台亭
3. 424凉亭
4. 荧光棒凉亭
5. 入口凉亭

Deck over a Roman Site in Cartagena

卡塔赫纳罗马遗迹平台

Completion date: 2011 **Location:** Cartagena, Spain **Designer:** Amann-Canovas-Mauri – Atxu Amann, Andrés Cánovas, Nicolás Maruri **Photographer:** David Frutos **Client:** Cartagena Puerto de Culturas

竣工时间：2011年 项目地点：西班牙，卡塔赫纳 设计师：阿曼-卡诺瓦斯-毛利——艾特修·阿曼、安德烈·卡诺瓦斯、尼古拉斯·玛鲁里 摄影师：大卫·弗鲁托斯 委托人：卡塔赫纳文化港口

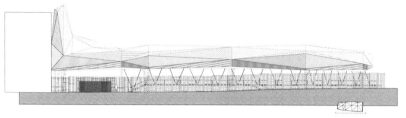

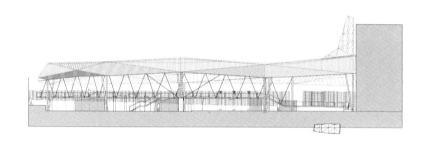

The building is essentially a cover protecting the remains of a Roman assembly (thermal baths, forum and domus) in the archaeological site of Molinete Park in Cartagena, Spain.

This cover is certainly another piece in the urban area of Cartagena whose main architectural challenge is to reconcile very different architectures, from the Roman times, passing through baroque to contemporary architectures, making them vibrate together in the neighborhood. It is a transition element, between very different city conditions, in size and structure, from the dense city centre to the slope park.

The primary goal of the project is to respect the existing remains, using a long-span structure, which requires the least amount of support for lifting the cover. The intervention unifies all the remains in a single space, allowing a continuous perception of the whole site. The cover also generates a new urban façade in the partition wall.

The project also pursues a sense of lightness and is conceived as an element that allows light. The inner layer is built with a modular system of corrugated multiwall translucent polycarbonate sheets. The outer layer, constructed with perforated steel plates, qualifies the incidence of light and gives a uniform exterior appearance.

Besides to the steel structure, the project proposes an elevated walkway parallel to the street. It is a very light structure hanging from the steel beams. Conceived as a glass box, with a faceted, partially visible geometry, it builds the street façade and allows a view of the ruins from three metres height. It is also accessible for disabled visitors. This high path permits an overall vision of the Roman remains.

从本质上来说，建筑是保护一处位于西班牙卡塔赫纳斗牛花招公园的罗马考古遗迹（浴场、论坛和住所）的顶盖。顶盖是卡塔赫纳市区一处显著的风景，它面临的主要建筑挑战是一系列建筑（从罗马时代、巴洛克时期到当代）统一在一起，让它们在街区中活跃起来。它是不同城市环境中的过渡元素，从规模到结构，从密集的城市中心到坡地公园。

项目的主要目标是利用大跨度结构来体现对原有的遗迹的尊重，而这个结构必须采用最少的支柱。设计将所有遗迹统一起来，在场地上形成了整体感。顶盖同时还通过隔断墙打造了一个全新的城市立面。

项目追求明亮感，被设计成一件透光元素。内层结构采用了多层波纹半透明聚碳酸酯板。外层由穿孔钢板构造，允许光线射入，为建筑打造了统一的外观。

除了钢结构，项目还设计与街道平行的高架走道。走道从钢梁上悬吊而下，十分轻盈。项目被设计成一个棱角分明的玻璃盒子，它打造了街道外立面，让人们从三米高的地方看到遗迹。项目同样设计了无障碍通道，高架走道让人们对罗马遗迹有了全面的了解。

1. The site pavilion in shining light
2. Façade structure steel tube
3. A bird's eye view of the framework of the site

1. 灯火辉煌的遗址亭
2. 外立面结构钢管
3. 俯视遗址区的架构

4. Visitor area
5. The site under the glass steel structure

4. 游客参观区
5. 玻璃钢结构下的遗址区

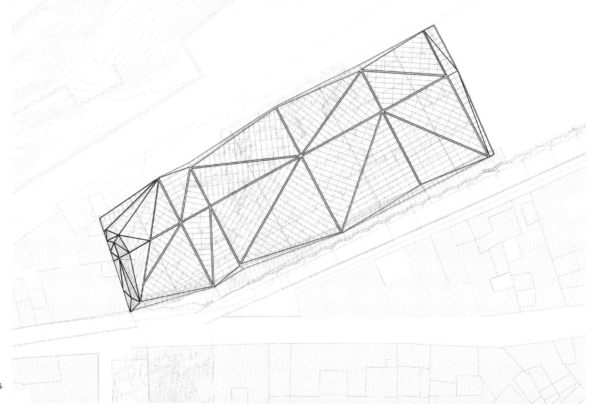

Plant truss
底部桁架

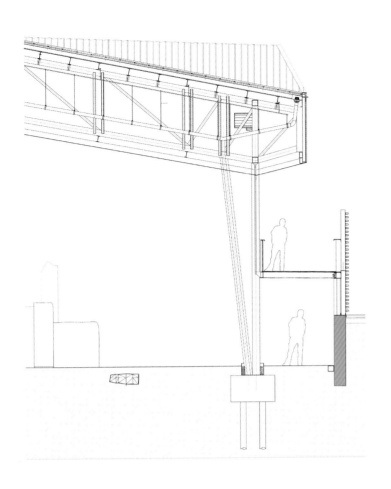
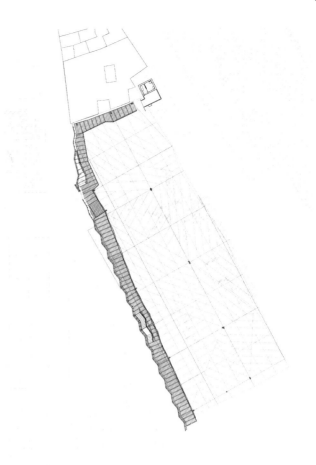

Gateway-plant
入口底座

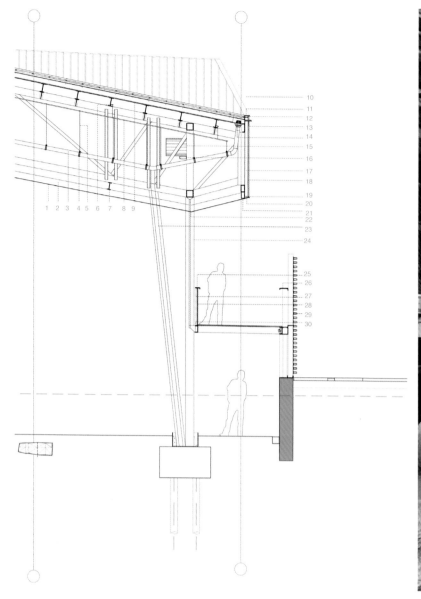

Deck over a Roman Site in Cartagen
1. Perforated steel plates. Dimensions 100x200cm, 1mm thickness (40% perforation)
2. Steel structure 1808100x60mm
3. Steel sewer d: 110mm, thickness: 0.6mm, 3.00m length
4. Steel clamp d: 110mm
5. Steel rods for hanging the plumbing net
6. Steel angle frame length: 80mm
7. Steel structure IPE200
8. Steel rails for hanging the plumbing net
9. Steel gutter, dimensions: 350x120mm, thickness: 0.8mm
10. Perforated steel plate (60% perforation)
11. Folded steel sheet dimensions: 30x140mm
12. Facade structure steel tube dimension: 100x50mm, thickness: 5mm
13. Steel tube for avoiding overflows
14. Galvanised steel drain
15. Lighting system
16. Perforated steel sheet for carrying cables
17. Facade structure IPE100
18. Perforated sheet plate wide: 1,090mm (40% perforation)
19. Z shape steel sheet. Length: 50mm thickness: 4mm
20. L shape steel sheet. Length: 50mm thickness: 4mm
21. Tubular structure dimensiong: 180x100x60
22. Steel sheet diameter: 160mm, thickness: 5mm
23. Lacquered steel pillar diameter: 110mm
24. Lacquered steel pillar
25. Banister: Lacquered folded steel sheet. Dimensions: 140x40mm, thickness: 10mm
26. Banister: Lacquered folded steel sheet. Dimensions: 250x40mm, thickness: 10mm
27. Steel structure for the banister 80mm wide
28. Wood floor
29. Cast acrylic sheets thickness: 10mm, 150mm wide
30. Steel structure UPN200

卡塔赫纳罗马遗迹平台
1. 穿孔钢板，尺寸100x200cm，1mm厚（40%穿孔）
2. 钢结构1808100x60mm
3. 钢污水管，直径110mm，厚0.6mm，长3.00m
4. 钢架：直径110mm
5. 悬挂水管网线的钢条
6. 角钢框架，80mm长
7. 钢结构IPE200
8. 悬挂水管网线的钢轨
9. 钢槽，尺寸350x120mm，厚度0.8mm
10. 穿孔钢板（60%穿孔）
11. 褶皱钢片，尺寸30x140mm
12. 外立面结构钢管，尺寸100x50mm，厚5mm
13. 防止溢流的钢管
14. 镀锌钢排水管
15. 照明系统
16. 承载缆索的穿孔钢板
17. 外立面结构IPE100
18. 传龙板，宽1,090mm（40%穿孔）
19. Z形钢片，长50mm，厚4mm
20. L形钢片，长50mm，厚4mm
21. 管状结构，尺寸180x100x60
22. 钢片，直径：160mm，厚5mm
23. 喷漆钢柱，直径110mm
24. 喷漆钢柱
25. 栏杆：喷漆褶皱钢片，尺寸140x40mm，厚10mm
26. 栏杆：喷漆褶皱钢片，尺寸140x40mm，厚10mm
27. 栏杆钢结构，厚80mm
28. 木地板
29. 浇铸亚克力板，厚10mm，宽150mm
30. 钢结构UPN200

6. A perfect transition from old Rome to modern world
7. Stairs of the raised visitor area

6. 从古罗马到现代的完美过渡
7. 高架人行参观区的楼梯

Vinaros Sea Pavilion

韦纳洛斯海滨凉亭

Completion date: 2009 **Location:** Castellon, Spain **Site area:** 360 sqm **Designer:** Guallart Architects
Photographer: Adriá Goula

竣工时间：2009年 项目地点：西班牙，卡斯特伦 占地面积：360平方米 设计师：加拉尔特建筑事务所 摄影师：阿德利亚·古拉

The Town Council of Vinaròs held a competition for the design and construction of a leisure area at the north end of the town's Mediterranean seafront promenade.

The architects decided to develop a parametric arboreal system in which all of the units were self-similar: both the structures that rest on the ground and those that rise up to expel fumes or take in light.

The structures are of 3mm painted galvanized steel, thin enough to be easily cut and folded to create a continuous hollow structural element. The openings in the structure are filled with glazed and opaque surfaces, on which the LED lights are mounted.

All of the structures have a hexagonal pattern that is deformed on a regular basis on the side facing the coast.

The process of fabricating these structures, involving the plasma cutting of some 1,736 different pieces, was an opportunity to experiment with the use of advanced manufacturing systems in which the architect directly produces the components of the project using parametric design and scripting programmes.

韦纳洛斯镇议会举办了一次设计竞赛，准备在城镇北端地中海滨海大道上建设一片休闲区。
建筑师决定开发一个系列相似的树冠系统，由地面上结构和上方的排烟和透光结构组成。
建筑结构由3毫米厚的涂漆镀锌钢板组成。薄板便于切割和折叠，以打造连续的中空结构元素。结构上的开口上装配这种玻璃和不透明表面，上方装有LED灯。
所有结构都采用了六边形图案，坐落在朝向海岸的底座上。
这些结构的制造过程包括1,736块不同构件的等离子切割，采用先进的制造系统，让建筑师能够利用参数设计和脚本处理直接参与到项目元件的制造过程当中。

Glass pavilion
玻璃凉亭

1. Leisure area at the north end of the Mediterranean
2. All of the Pavilions' units were self-similar.
3. Access of the pavilion

1. 地中海北端的休闲区
2. 所有凉亭单元都相互类似
3. 凉亭入口

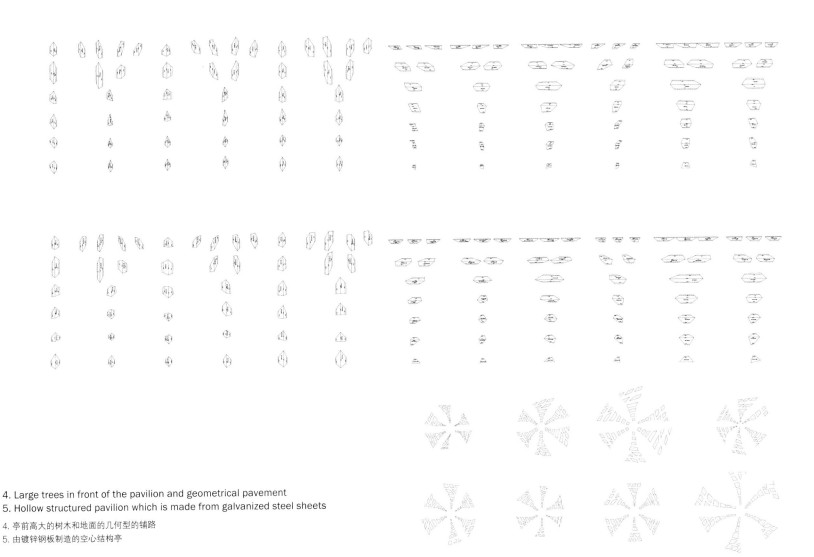

Glass development
玻璃开发

4. Large trees in front of the pavilion and geometrical pavement
5. Hollow structured pavilion which is made from galvanized steel sheets
4. 亭前高大的树木和地面的几何型的铺路
5. 由镀锌钢板制造的空心结构亭

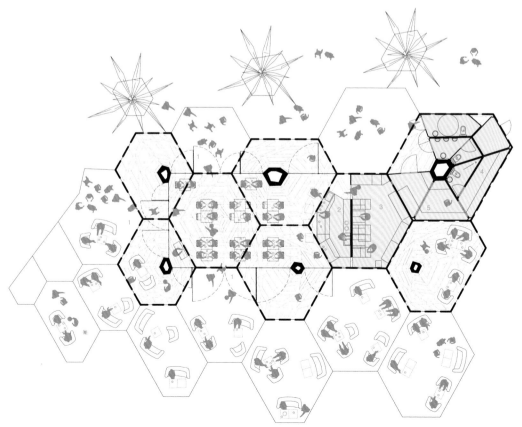

1. Access
2. Bar
3. Kitchen
4. Store
5. Office
1. 入口
2. 酒吧
3. 厨房
4. 仓库
5. 办公室

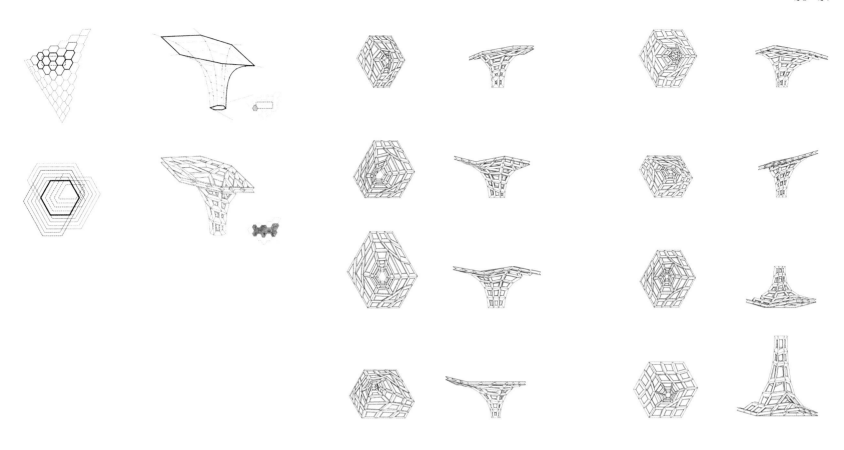

6. LED lights are installed in the middle of the square.
7. All of the structures have a hexagonal pattern.
8. People gathering in the red LED lighting

6. LED灯装在镀锌板结构中的四边形中间
7. 所有结构都采用了六边形造型
8. 红色LED灯下聚会的人们

Iron Bark Ridge Park Environment Centre

澳洲橡木山公园环境中心

Completion date: 2010 **Location:** Sydney, Australia **Designer:** hungerford+edmunds **Photographer:** Simon Wood

竣工时间：2010年 项目地点：澳大利亚，悉尼 设计师：亨格福特+艾德蒙斯 摄影师：西蒙·伍德

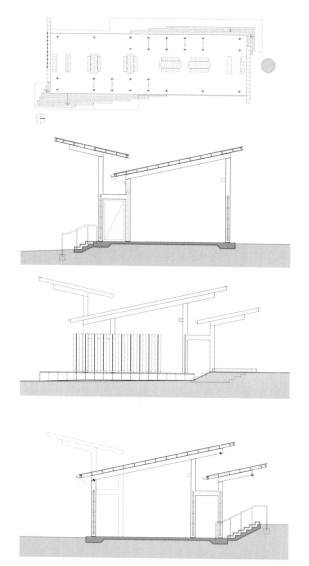

The Environment Centre at Iron Bark Ridge Park is a covered, open-air pavilion; it provides a classroom and meeting venue for local schools and community groups; and picnic shelter for the general public.

Iron Bark Ridge Park is connected by cycleways to a larger network of parks, and open spaces that have been designed by Oculus as part of the New Rouse Hill. The New Rouse Hill is an award winning new community, located in Sydney, Australia and is being delivered as a joint venture between Lend Lease and the GPT Group, in active partnership with Landcom and the NSW Department of Planning and Infrastructure.

The Environment Centre is located against the trees at the edge of a gently sloping meadow. The site gradually falls 1.5m from south to north, and the pavilion accommodates this grade change diagonally, creating amphitheatre seating.

With no walls, the roof is the elevation. This is articulated by employing three different planes, with varying height and presentation. While the portal frames share pitch, and rhythm, the triple roof lends subtle variations in form and connection. The galvanised finish of the primary supporting structure is clearly legible against the intense colours of the soffits, which in turn contrast with the sky and tree canopy.

The building typology, an open air meeting venue, is found throughout Sydney, especially in the western suburbs and is one that is always well used and enjoyed by the public. This building adds yet another space for the community to enjoy.

澳洲橡木山公园环境中心是一个露天凉亭，它为当地学校和社区群体提供了露天教室和集会场所，也是公众的野餐亭。

澳洲橡木山公园通过自行车道与奥库卢斯设计的新洛兹山社区更大的公园网络和开放空间相连。富有盛名的新洛兹山社区位于澳大利亚悉尼，由联盛集团和GPT集团联合经营，与兰德科姆和新南威尔士州规划和基建部紧密合作。

环境中心坐落在一片草坡的边缘，紧靠树林。场地由南至北高度差为1.5米，凉亭的对角线利用了这个高度差，创建了阶梯坐席。

凉亭没有墙壁，屋顶就是唯一的立面。它由三块高度和造型各不相同的面板铰接而成。门式钢架略微倾斜，具有韵律感，导致三重屋顶在造型和连接上有着微妙的不同。主要支承结构的镀锌包层在屋顶下表面强烈色彩的对比下，显得异常清晰。同时，下表面的色彩也与天空和树冠形成了鲜明对比。

这种建筑类型——露天集会场所——在悉尼随处可见，尤其是悉尼西郊，深受公众的喜爱。建筑为公众提供了一个全新的享乐场所。

1. The open-air pavilion is located in Iron Bark Ridge Park.
2. The pavilion is located against the trees at the edge of a gently sloping meadow.

1. 露天凉亭位于橡木山公园
2. 凉亭坐落在草地中，紧靠树林

3. The roof consists of three different planes, with varying height.
4. The seat's colour matches with the colour of the roof lower surface.
5. The triple roof lends subtle variations in form and connection.

3. 亭顶是三块高低不相同的面板
4. 坐席的色彩与屋顶下表面的色彩相同
5. 三重亭顶在造型和连接上有着微妙的不同

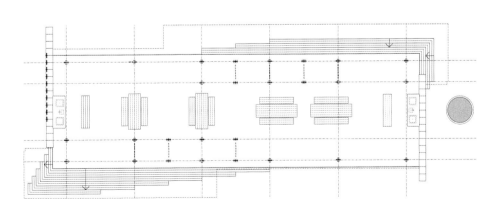

Ruth Lily Visitors Pavilion

卢斯·莉莉游客亭

Completion date: 2010 **Location:** Indianapolis, USA **Area:** 120 sqm **Designer:** Marlon Blackwell Architect **Photographer:** Timothy Hursley, Marlon Blackwell Architect staff **Client:** Indianapolis Museum of Art

竣工时间：2010年 项目地点：美国，印第安纳波利斯 项目面积：120平方米 设计师：马龙·布莱克韦尔建筑事务所 摄影师：蒂莫西·赫斯利、马龙·布莱克韦尔建筑事务所员工 委托人：印第安纳波利斯艺术博物馆

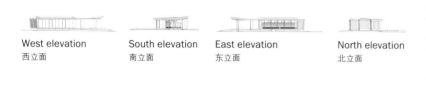

| West elevation | South elevation | East elevation | North elevation |
| 西立面 | 南立面 | 东立面 | 北立面 |

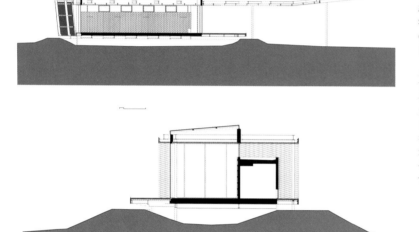

The Ruth Lilly Visitors Pavilion is the result of a studied relationship between building, land and art, and serves as both a threshold and a destination within the 100 Acres Art & Nature Park at the Indianapolis Museum of Art. An ipe screen lines a steel exoskeleton forming deck, wall and canopy, wrapping programmatic elements. The pavilion is constructed to touch the landscape lightly and allow for the free flow of rain and filtered sunlight through the structure.

The interior space is surrounded by glass on three sides in addition to the skylight ceiling above, allowing visitors to maintain a powerful connection to the natural world around them. The building provides a versatile gathering and education space, restrooms, and emergency services. Carefully sited and in the woods of 100 Acres, the ADA accessible building is accessed by the park's network of pedestrian landscape journeys.

The Visitors Pavilion will be LEED certified, with careful attention paid to environmental sensitivity and energy efficiency throughout design and construction. Water saving fixtures are fed by on-site well water and a geothermal system provides heating and cooling. The Visitors Pavilion and surrounding landforms are carefully designed to inhabit the floodplain environment of 100 Acres, allowing occasional floodwaters to pass around and beneath the structure.

卢斯·莉莉游客亭的设计深刻研究了建筑、土地和艺术之间的关系，是印第安纳波利斯艺术博物馆100英亩艺术与自然园区的门户。重蚁木屏与钢铁外框架组成了平台、墙壁和天篷，将项目元素包裹起来。凉亭与景观轻触，让雨水可以自由流动，同时木屏结构还能过滤阳光。

室内空间三面由玻璃包围，上方是带天窗的天花板，让游客与周边的自然环境能够亲密接触。建筑提供了多功能集会和教育场所、洗手间和急救服务。这座无功能障碍建筑巧妙地坐落在树林之中，可以通过公园的人行网络进入。

游客亭即将通过绿色建筑认证，在设计和建造过程中十分注重环境敏感度和能源效率。节水设施由现场的井水提供用水，而地热系统则提供了供暖和制冷。游客亭和周边的地貌十分切合公园泛滥平原的环境，让偶发的洪水从四周绕过或从下方流过。

1. Ruth Lily Visitors Pavilion is the result of a studied relationship between building, land and art.
2. The ipe screen combines the pavilion to the surrounding trees.
3. Path to the visitors pavilion

1. 卢斯·莉莉游客亭的设计完美的体现了建筑、土地和艺术之间的关系
2. 木屏的设计与周围的树木融为一体
3. 通向游客亭的小径

Awards:
2011 Arkansas AIA Honour Award
2011 Indianapolis Sustainability Awards IndyGo Green – Land
2011 Shortlisted for World Architecture Festival Award in the Display and Visitor centres category

获奖情况：
2011美国建筑师协会阿肯色氛围荣誉奖
2011印第安纳波利斯可持续设计奖
2011世界建筑节奖展示与游客中心类提名

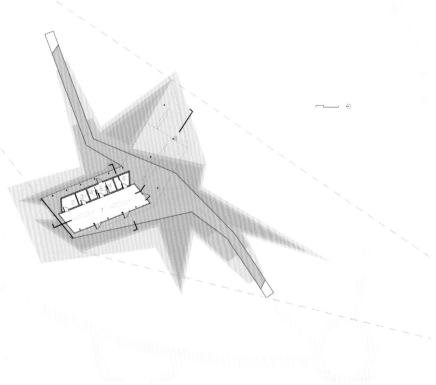

1. Multi-purpose
2. Office
3. Kitchen
4. Restroom
5. Vestibule
6. Storage + mechanical
7. Deck
8. Terrace

1. 多功能区
2. 办公室
3. 厨房
4. 洗手间
5. 门廊
6. 仓库+机械室
7. 平台
8. 露台

4. Interior space with glazed walls
5. Functional area

4. 带玻璃幕墙的室内空间
5. 多功能区

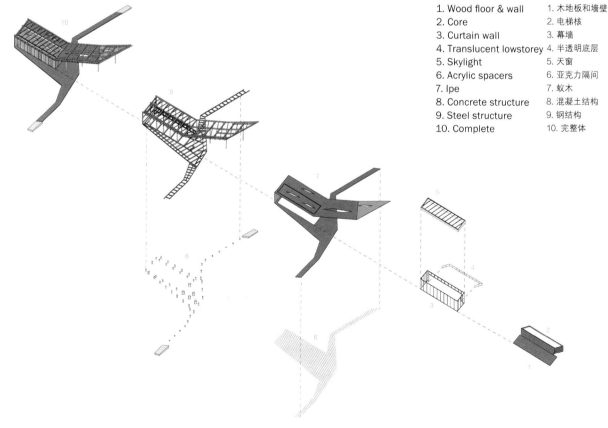

1. Wood floor & wall
2. Core
3. Curtain wall
4. Translucent lowstorey
5. Skylight
6. Acrylic spacers
7. Ipe
8. Concrete structure
9. Steel structure
10. Complete

1. 木地板和墙壁
2. 电梯核
3. 幕墙
4. 半透明底层
5. 天窗
6. 亚克力隔间
7. 蚁木
8. 混凝土结构
9. 钢结构
10. 完整体

Landscape Restaurant/Café

景区餐饮建筑

For the protection of the scenic spot, this kind of buildings needs special permission. In this chapter, sightseeing-type buildings contain three characteristics:
- Buildings belong to a park, or scenic spot
- Providing services for tourists
- Engaged in the catering business, including restaurant, bar and small cafeteria.

1. Landscape & Architecture

1.1 Nature and environment

The sightseeing-type building's flat shape, type, wall, door, window, ceiling and floor are all limited by the natural situation including appropriateness, sunlight, rain and snow, wind speed, wind direction as well as typography, geological conditions and earthquake intensity. Meantime, the building's floor plan layout, type, elevation façade, site layout and so on are limited by the surrounding architecture, road, afforestation. Separating nature from environment to design is unimaginable.

1.2 Relationship between sightseeing-type building function and architectural type

Different function requires different architectural space, while different architectural space forms the construction entity, which forms the variety of architectural appearance. Hence, different architectural types are formed. Generally speaking, an excellent architectural exterior outlook should fully reflect the requirements of interior space and the different characteristics of the building, which achieves the dialectical unity of form and content. Landscape building especially emphasises the design principle of both the architectural function and type.

1.3 Environment and sightseeing-type building

Landscape building can not only bring beauty to the environment, but also satisfy the needs for people's rest and cultural entertainment. The environment becomes more beautiful because of the decoration of architecture, which is more suitable for people's visiting and enjoying the view; and architecture exists for the surrounding landscape and serves for it. Landscape building is the combination of landscape and architecture, which comes from the admiration for natural scenery. Its architecture design should help to add the view as well as coordinate with the surrounding environment and try to make full use of the natural typography and landform, which satisfy the general

为保护景区的环境，景区中商业建筑的经营需得到特许经营许可。本章所阐述的景区小型商业建筑包含3个特征：
- 建筑位于风景区中
- 为游客提供服务
- 从事饮食经营活动的建筑物

1．建筑与景观

1.1 自然与环境

建筑物的平面形状、体型及墙体、门窗、屋顶、地面等围护结构都要受到自然条件，包括湿度、日照、雨雪、风速、风向等气候条件及地形、地质条件以及地震烈度等的限制和制约，同时建筑物的平面布置、体型、立面造型、场地布置等还要受到其周围建筑、道路、绿化等环境的限制，脱离自然与环境来做设计是难以想象的。

1.2 建筑功能与建筑造型的关系

不同的功能要求形成了不同的建筑空间，而不同的建筑空间所构成的建筑实体又形成建筑外形的变化，因而产生了不同类型的建筑造型。与此同时，建筑的造型又反映出建筑的性质、类型。形式服从功能是建筑设计遵循的原则。一般一个优秀的建筑外部形象必然要充分反映出室内空间的要求和建筑物的不同性格特征，达到形式与内容的辩证统一，景观建筑尤其强调建筑功能和建筑造型并重的设计原则。

1.3 环境与建筑

景观建筑既能美化环境，又能满足人们的休息和文化娱乐生活需要。环境因为建筑的点缀而更加优美，更适合人们游玩、观赏的需要；而建筑是因周围的景观而存在，为环境服务。景观建筑是景观与建筑的结合体，是出自对自然景色的向往，其建筑设计要有助于增添景色，并与周围环

requirement and the need for enjoying the view, thus, a harmonised space with the nature is created.

Landscape building should define the architecture nature and orientation first in design and take the architectural function, material, environment and culture into consideration. Second, the rationality of the layout should also be considered. The distribution of different buildings can be decided according to the visitors' possible actions. The building interior should actually be aware of the locations, passageways and layout of the different functions.

The new, user-friendly facilities, such as the Pavillon Madeleine Restaurant, the seating, lights, masts, bridges bus stop and path borders all feature corten steel, as a recurring theme throughout the park. The materials and function of the façade were given particular attention as a result of the requirements of the building façades and outdoor furniture, i.e. ensuring that they are resistant to vandalism and that they blend in well in the green surroundings. The use of corten steel highlights the cultural heritage of the steel industry in the south of Luxembourg. The material is particularly striking due to its sustainability during use and is fully recyclable. It boasts a unique and versatile range of applications. Figure 1

境相协调，充分利用自然地形、地貌的条件，满足环境的整体要求和游人欣赏自然风景的需要，从而创造和大自然相协调的空间。

景观建筑在设计上应先确定建筑性质和定位，考虑建筑的功能、材料、环境、文化等因素。其次，要考虑建筑布局的合理性，根据游客可能的行为来确定不同性质建筑的分布，建筑内部实际更要注意各功能区之间的位置关系、通道及布局等。

玛德琳餐厅的座椅、照明、船桅、汽车站和小路的边缘设计都使用了耐候钢，从而贯穿了公园的主题。由于建筑外墙和室外家具的特殊要求，建筑表面的材料和功能受到了特别的关注，比如，要确保它们可以抵御人为破坏，并且与周围的绿色环境相协调。耐候钢的使用彰显出卢森堡南部钢铁文化遗址的特点。由于其在使用中的可持续性和可完全再回收的特性，耐候钢这种材料尤其引人注意。它以其独特而又广泛的使用性能著称。如图1

Figure 7: Pavilion Madeleine Restaurant, see page 68
图1 玛德琳餐厅，参见68页

Landscape Restaurant/Café

Figure 2. The small-sized Yellow Treehouse Restaurant, together with the surrounding trees, has been integrated into the local landform, making the building and the environmental a perfect match.
图2 黄色树屋餐厅的体量很小，与周围的树林融为一体，充分与该地地貌相结合，使建筑与环境空间配对。

2. Serving sightseeing-type building

Landscape building can be divided into leisure architecture, serving architecture, cultural entertainment architecture, and manageable architecture according to its usage functions. Serving Landscape building is a significant factor of the landscape environment, which provides visitors with massages, education, food and drink, rest and so on, including restaurant, bar, small cafeteria, marinas, exhibition room, ticket office, garden toilet and other practical buildings. These buildings' sizes are small, but close to people, combing the usage function with the art of landscape. It proceeds from the whole situation in terms of the location, direction, pattern, size, height, style and colour, design in coordination with each other spatially, and make full use of the typography and landform, which preserve the natural environment and not destroy the existing landscape, thus, the natural environment and landscape is harmonised. The design should suit one's measures to local conditions, give environment to its full advantages and create a rich space. Serving landscape building should create certain conditions for visitors to enjoy the scenery. The size and volume should be batted together with creating a good field of vision. The height of the building should be subject to the landscape, at the same time, overall idea is required. It makes no sense to have one without another. It is necessary to create a new landscape effect, but also take the surrounding scenery into consideration.

Designed by Pacific Environments Architects, the concept of Yellow Treehouse Restaurant is driven by the 'enchanted' site which is raised above an open meadow and meandering stream on the edge of the woods. The tree-house concept is reminiscent of childhood dreams and playtime, fairy stories of enchantment and imagination. It's inspired through many forms found in nature: the chrysalis/cocoon protecting the emerging butterfly/moth, perhaps an onion/garlic clove form hung out to dry. It is also seen as a lantern, a beacon at night that simply glows yet during the day it might be a semi camouflaged growth, or a tree fort that provides an outlook and that offers refuge. The plan form also has loose similarities to a sea shell with the open ends spiraling to the centre. Figure 2

2. 服务性建筑

服务性景观建筑是景观环境的重要组成要素，为游人提供信息、教育、饮食、休息等服务，包括餐厅、酒吧、小卖部、游船码头、展览室、售票房、园厕等具有实用价值的建筑。这些建筑虽然体量不大，但与人们密切相关，融使用功能与艺术造景为一体，在位置、朝向、造型、体量、高度、风格、色彩等上都要从全局的总体布置出发，在空间上相互协调和呼应，充分与地形、地貌相结合，保持自然环境，不损害原有景观，并与自然环境与景观统一协调。在设计上应因地制宜，发挥环境优势，创造丰富空间。服务性景观建筑要为游人赏景创造一定的条件，应反复推敲体型、体量，也要创造良好的视野，建筑高度应服从景观需要，同时要有全局观念，不能顾此失彼。既要创造新的景观效果，也要顾及到周边的景色。

由太平洋环境建筑事务所设计的黄色树屋餐厅位于一片开敞的草甸上，树林边缘是蜿蜒的小溪，设计概念正是受到了"魔法场地"的启发。树屋理念让人们回忆起童年梦想和玩乐时间、魔法童话和想象。它受到了大自然中多种形态的启发——保护蝴蝶和蛾子的虫茧、挂在外面晾干的洋葱头和大蒜瓣等。它还可以被看做一盏灯笼、一座夜间的灯塔、一座树上侦查堡垒。同时，建筑造型还与中空螺旋海螺类似。选定的场地和树必须符合一系列功能要求——树屋需要同时容纳18名客人、服务员和吧台；通过辅助灯光获得正确的摄像角度，适合拍摄广告、网络摄像和拍摄剧照；拥有一览无余的山谷和场地入口景色；坚固的结构。最终所选择的树木较为高大，位于一处陡坡上，进一步凸显了树的高度。厨房/备餐设施和洗手间都设在地面。如图2

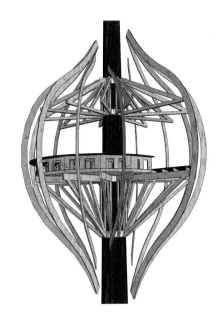

Figure 3. Construction of tree-house

The Architectural component embodies a simple oval form wrapped 'organically' around the trunk and structurally tied at top and bottom, with a circular plan that is split apart on the axis with the rear floor portion raised (Figure 3). This allows the approach from the rear via a playful tree-top walkway experience, slipping inside the exposed face of the pod and being enchanted by the juxtaposition of being in an enclosed space that is also quite 'open' and permeable to the treetop views. There is also a 'Juliet' deck opposite the entrance that looks down the valley.

The scale and form of the tree-house creates a memorable statement without dominating its setting. While it's natural 'organic' form sits comfortably, the rhythm of the various materials retains its strong architectural statement. The verticality of the fins mimics the verticality of the redwoods and enables the building to naturally 'blend' into its setting, as though it were a natural growth.

3. Tourist catering building

Catering building is a place for dining and resting for visitors, including restaurant, bar and small cafeteria. This kind of building is built in the park or in the scenic spot, thus, the location and design of the building is appropriate or not has close relations with the visitors and adding beauty to the scenic spot and park.

3.1 The choice of location
- Should be located in people concentrated spots nearby, and pay attention that the traffic is convenient, easily accessible, convenient management and supply of goods and materials. Avoid too remote or centralised facilities area;
- Avoid from the various sources of pollution, and meet the health protection standards;
- Good lighting and ventilation conditions;
- A beautiful environment, make full use of the surrounding environment and combining with nature;
- To provide necessary conditions for building function zoning, the arrangement of the entrance, the layout of outdoor venues.

3.2 Design points
The construction of the indoor and outdoor space should blend with each other, enriching architectural space level and promoting the interaction between architectural and environment. The tourism season changes can also be adjusted to use outdoor

Landscape Restaurant/Café

space. For example, there are more people in spring and summer, outdoor space can be used to regulate the crowd, setting flowers, pavilion, corridor construction and so on, which can not only shade sunlight and rain, but also for leisure and chat.

Business premises should be selected in the best solar azimuth, as far as possible to avoid the cold winter wind invasion and the exposure in summer. Pay attention to ventilation conditions, creating a good view. The interior facility should be arranged reasonable and unobstructed, avoiding the interference between the flow lines. The desks and chairs should be arranged according to the principle of human engineering, so as to bring a more comfortable environment.

3.3 Ecological design

Any design that coordinated with ecological processes, minimise the damage on the environment to achieve the minimum design form is called ecological design. This coordination means the design respects for diversity, reduce the deprivation of resources, keep nutrition and water, and maintain plant habitat and animal habitat quality, in order to contribute to the improvement of the living environment and ecological system health. Ecological design provides us with a unified framework, helping us to reexamine the landscape architectural design and people's daily life style and behaviour. Simply speaking, the ecology design is the process of natural effective adaptation and combination, which needs to have a comprehensive measure of the design approach on environment impact.

In a dune nature reserve rises a new restaurant: Aan Zee. It is situated on a breathtaking, rather surreal location: wedged between the dune nature reserve 'Voornes duin' and the Rotterdam harbour extension 'Maasvlakte', which fills the horizons with heavy industry and rare birds simultaneously. Emma Architects was responsible for the concept, design and realisation of the building and its immediate surroundings, its interiors and the restaurant concept. Figure 4

It is a new restaurant in many ways. It is autarkic in its energy supply, water supply and waste water treatment. It uses sun, wind, wood fire, natural convection and the earth itself as sources for energy, heating, cooling, ventilation, and waste water management. The cook works with a wood burn stove only, and uses ingredients from local farmers and fishermen. The building itself can be fully dismantled and is re-usable both as a

多利用室外空间调节人流，设置花架，亭、廊等建筑，既可遮阳又可避雨，还可休闲、聊天。

营业用房应选择在最好的日照方位，尽量避免冬天的寒风侵袭夏日的炎日照射，注意通风条件，创造良好的景观视线。内部设施要合理布置，通畅，避免各流线间的干扰，桌椅布置应该符合人体工程学原理，以便为人们带来更加舒适的环境。

3.3 生态设计

任何与生态过程相协调，尽量使其对环境的破坏影响达到最小的设计形式都称为生态设计，这种协调意味着设计尊重物种多样性，减少对资源的剥夺，保持营养和水循环，维持植物环境和动物栖息地的质量，以有助于改善人居环境及生态系统的健康。生态设计为我们提供一个统一的框架，帮助我们重新审视景观建筑设计以及改变人们日常生活方式和行为。简单地说，生态设计是对自然过程的有效适应及结合，它需要对设计途径给环境带来的冲击进行全面的衡量。

天然的沙堆中建造起一座新餐厅——"克安吉（Aan Zee）"。这个餐厅坐落于令人惊讶的超现实地理位置上：楔形插在天然沙堆"Voornes duin"和鹿特丹港口延伸区域"Maasvlakte"之间，这里到处都是重工业，人迹罕至。艾玛建筑事务所负责其理念、设计和建成，同时负责其周边设施、室内和餐厅的设计构思。如图4

从很多方面讲它都是一个新餐厅。能源供给，水供给和废水处理都可以自行解决。餐厅利用阳光、风、木柴、自然对流和土壤本身作为能源、热量、制冷、消毒和处理废水的资源。厨师做饭的工具是一个木制的炉子，并且从当地农民和渔民那里买来原料。建筑本身还可以全部拆开，所

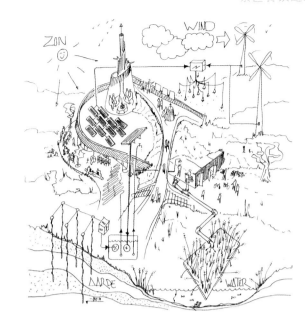

Figure 4. Site plan, Aan Zee Restaurant, Emma Architects
图4 克安吉餐厅总体规划设计，艾玛建筑事务所

building as on a level of materials.

A wooden wall curves up from the dunes, it curles up to form a watchtower. The wood is layered and in fixed in overlapping small elements on the façade and roofline. The untreated local wood has been cooked to withstand the elements without chemical treatments, varnish or paint. It will change colour over the years, and differently in each direction. The wind and sun will colour this building, making it a micro climate report in the sand. This wooden wall encloses the restaurant space, together with an all glass curved façade, aimed at the astonishing views towards the sea. The back of house is made up from concrete standard issue basements, buried under the sand, covered with sea. Figure 5

以无论从建筑还是原材料的层面上讲都是可以再利用的。

木质的墙壁从沙堆中慢慢上升，弯曲蜷缩形成一个观景塔楼元素。木质材料形成不同的叠层，同时包含不同的小单元。这些取材于的木材未经过任何化学处理，随着时间的流逝，色彩也会发生变化。一个玻璃表皮的融入确保了灯光和视野。这个木墙围住了餐厅的空间，还有都是用玻璃包住的弯曲表面，这样设计的目的是为了有一个更好的角度来看海。屋子的后面建有标准的混凝土地下室，埋于沙子之下，上面有海水覆盖。如图5

Figure 5. Full view of Aan Zee Restaurant, Emma Architects
图5 克安吉餐厅全景图

天门山"山之港"临江餐厅 Riverside Restaurant

Completion date: 2011 **Location:** China, Guilin **Designer:** Liu Chongxiao **Photographer:** Deng Xixun, Liu Chongxiao, He Rong, Song Yan

竣工时间：2011 项目地点：中国，桂林 设计师：刘崇霄 摄影师：邓熙勋、刘崇霄、何蓉、宋彦

Riverside Restaurant is fantastic for its small scale and pure function which make it fused into the surroundings in a semi-transparent way as well. The solidness of structure and material and the illusory feeling from visual and tactual aspect are combined together in a surprising way. Sightseeing is not to be observed nor the architecture a shelter; they permeate into each other subtlety and the boundary between them becomes obscure.

Rather than submerged into the surroundings completely, the design aims to redefine the features of the surroundings by bringing architecture in and at the same time to dug out the potential of the surrounding background. What is more important, there is no clear difference between exterior and interior design. The starting point is the interior.

From the outside, once can feel the fusion of the architectural exterior and the surrounding bamboo forest, mountains and interior materials; once inside, one can enjoy the feeling of being enveloped by architectural materials and sightseeing under certain light. Such effect is reproduced by the reinterpretation of exterior scenic views through architectural design.

临江餐厅是一个奇妙的小项目，正是空间的小和功能的纯粹，为它以"半透明"的身份融进周边环境提供了条件。设计出乎意料地把结构与材料的真实性和视觉与触觉的虚幻感结合在一起。景色，已经不再是被观察的对象。建筑，已经不再是遮蔽所。它们以某种微妙的方式互相渗透着，建筑和环境之间的边界变得模糊起来。

与完全消隐不同，餐厅的设计期待通过建筑的嵌入对周遭环境的特征加以重新定义，揭示出原有环境中潜藏的某些活力。此外，非常重要的一点是，整栋建筑没有建筑设计与室内设计的分别，它是从内部开始并由内而外开始思考的。

人们在外部的某些角度观察建筑，通过建筑不同界面虚实关系的组合，会感受到某种建筑外表面与竹林、山体，室内材料质感的混合，而进入室内则将体会到某种光线下被建筑材质和风景组合包围的感受，那是借助建筑界面对外部景色的重新编辑产生的。

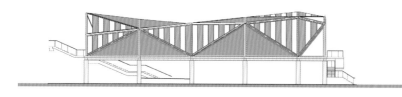

1. Elevation facing the river
2. The restaurant inserts the building into the surroundings.

1. 江景立面
2. 餐厅的设计使建筑嵌入对周遭的环境

3. The orange roof stands out of the green hills and rivers.
4. Entrance vestibule
5. Details of the architecture
6. A close shot of the façade

3. 橘色的屋顶在青山碧水中格外鲜明
4. 入口通廊
5. 建筑细部
6. 近景立面

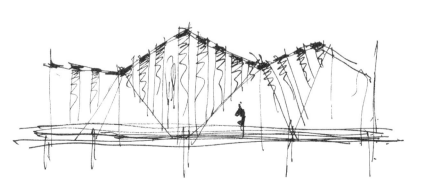
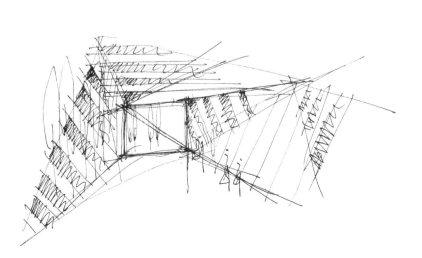

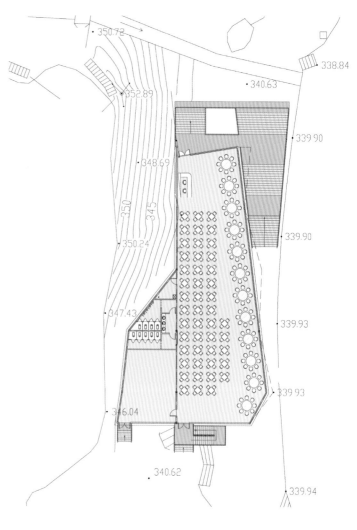

Plan
平面图

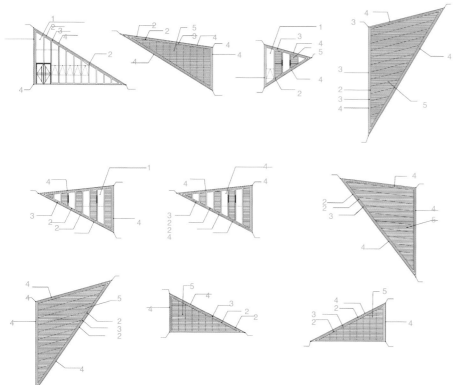

Details of the curtain wall
幕墙细节

1. 6+9A+6 Low-E hollow and transparent toughened glass
2. Perfluocarbon sprayed dark grey
3. Perfluocarbon sprayed light grey
4. Central line of structure menber bars
5. Plastic plate

1. 6+9A+6-E
   中空钢化透明玻璃
2. 氧碳喷涂深灰色
3. 氧碳喷涂浅灰色
4. 结构杆件中心线
5. 塑木板

7. Main entrance
8. Dining hall
9. Skylights and side windows

7. 建筑入口
8. 用餐大厅
9. 天窗与侧窗

HOTO餐厅 HOTO FUDO

Completion date: 2009 **Location:** Yamanashi, JAPAN **Site area:** 2,499 sqm **Building area:** 734 sqm
Building height: 7.46 m **Designer:** TAKESHI HOSAKA **Photographer:** Koji Fujii/Nacasa & Partners Inc.

竣工时间：2009年 项目地点：日本，山梨县 建筑面积：34平方米 建筑高度：7.46米 设计师：保坂健 摄影师：藤井浩司/纳卡萨摄影

The project was planned on the site with Mt. Fuji rising closely in the south and the two sides facing the trunk roads.

This building seems to belong to such nature objects as mountains and clouds. It is made from soft geometry, which will not arise from the figures like quadrangles and circles. By continuously operating innumerable polygon mesh points, the architects have determined the shape that clears the conditions such as the consistency as shell construction and the undulations that ward off rainwater in spite of its free geometry. The RC shell with cubic surfaces creates such spaces as 530 square metres of seats, 140 square metres of kitchens, and 50 square metres of rest rooms, in such a manner that it envelops and opens them.

This building has no air conditioners. It is open to the air at most seasons, and people have a meal in the air like outside air. The curved acrylic sliding door is closed only during the strong wind and the coldest season. Giving 60 mm thick urethane insulation to the outside of the RC shell and keeping a stable RC temperature secures a stable temperature environment for the building like inside and outside, and also reduces the deformation volume due to the temperature of RC to make the building last longer. For the lighting plan, the architects have determined such illumination as makes people simply feel changes in the evening light and does not make insects gather around the lights.

When it rains, rain comes in near windows and doors. In the spaces where rain does not come in, people enjoy the sound of raindrops. When it is foggy, the fog comes into the building. When it snows, it becomes a landscape buried in snow, and birds and animals will visit there. In this place like the middle between nature and art, people eat hoto rich in natural ingredients.

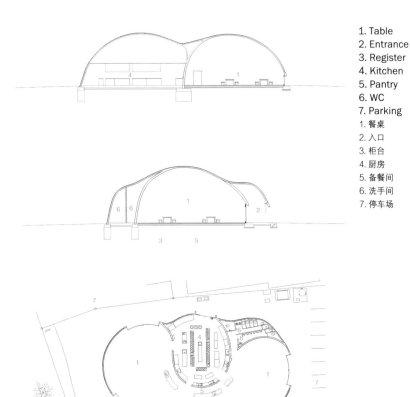

1. Table
2. Entrance
3. Register
4. Kitchen
5. Pantry
6. WC
7. Parking

1. 餐桌
2. 入口
3. 柜台
4. 厨房
5. 备餐间
6. 洗手间
7. 停车场

项目位于富士山南侧，两面朝向公路主干道。

建筑如此自然，似乎属于山峰和云彩。它由柔和的几何造型构成，没有简单的四方棱角和圆圈。建筑师通过数不清的连续多边形网格点来决定外壳结构的一致性；尽管造型自由，建筑的波浪造型依然能避开雨水。这个立体造型外壳所覆盖的空间包括：530平方米的坐席空间、140平方米的厨房和50平方米的洗手间。

建筑没有空调系统。大部分时节，它都对外开放，让人们仿佛在露天环境中就餐一样。弧形亚克力拉门只有在狂风和最寒冷的季节才会关闭。建筑外壳拥有60毫米的氨基甲酸乙酯隔热层，能够保持建筑恒温，并且减少结构变形，让建筑能够更加持久。在照明方面，建筑师希望人们能够简单地感受到夜间光，并保证昆虫不会在灯光四周聚集。

雨天，雨滴从门窗进入。在远离雨水的地方，人们可以倾听雨滴的声响。雾天，雾气会进入室内。雪天，建筑埋在雪里，形成了独特的景观，招来小鸟和动物。在这个介于自然与艺术之间的空间中，人们可以尽情享用天然成分制成的面条。

1-3. Arched acryl sliding door
4. The project was planned on the site with Mt. Fuji rising closely in the south.

1-3. 弧形亚克力拉门
4. 项目位于富士山南侧

* HOTO is traditional local noodle food.

* HOTO是一种当地传统面条。

5. The building looks like mountain or cloud.
6. When the door opens, people seem to have a meal in the air like outside air.
7. 530 square metres of seats

5. 餐厅的造型像山峰和云彩
6. 餐厅的门开放着,人们仿佛在露天环境中就餐
7. 座位区面积为530平方米

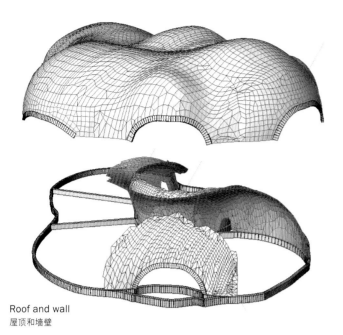

Roof and wall
屋顶和墙壁

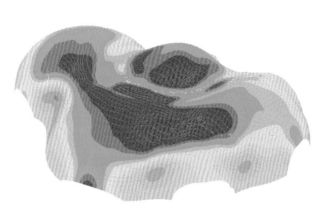

Element list: not PB1to 3000 not PAB
元素列表：not PB1to 3000 not PAB

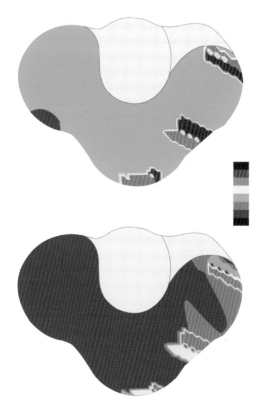

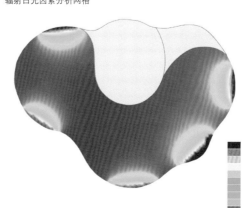

Analysis Grid RAD Daylight Factors
辐射日光因素分析网格

PMV distribution (16:00 in winter)
shell
1. Photocatalyst
2. Urethane coating waterproofing
3. GRC t=15mm
4. Insulation material Urethane t=60mm
5. RC shell t=100mm
6. Mortar t=15mm + Finish plaster t=2mm
7. Reinforcing rod unit

热环境综合评价指标（冬季16:00）
外壳
1. 光催化剂
2. 氨基甲酸乙酯防水涂层
3. 玻璃纤维增强水泥，厚15mm
4. 绝缘材料氨基甲酸乙酯，厚60mm
5. RC外壳，厚100mm
6. 砂浆（厚15mm）和石膏（厚2mm）
7. 钢筋

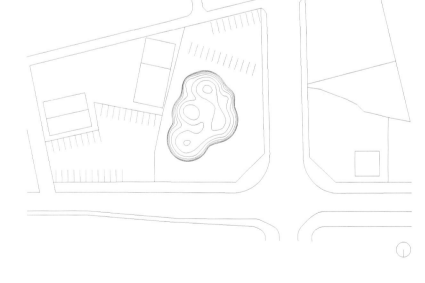

8. People could sense night light simply.
9-11. Interior space with free forms
12. People could enjoy Mt. Fuji's view in the restaurant.

8. 人们能够简单地感受到夜间光
9-11. 造型自由的室内空间
12. 餐厅内可以观赏富士山的美景

The Terrace View Café

露台咖啡厅

Completion date: 2009 **Location:** St. Louis, USA **Construction areas:** 307 sqm **Designer:** Studio|Durham Architects **Photographer:** Steve Hall, Hedrich Blessing Photographers, Chris Sauer/Christian Sauer Images

竣工时间：2009年 项目地点：美国，圣路易斯 建筑面积：307平方米 设计师：达勒姆建筑事务所 摄影师：斯蒂夫·霍尔、赫德瑞奇·布莱辛摄影、克里斯·萨奥尔/克里斯·萨奥尔图像

The Gateway Foundation hired Durham Architects to design a café and maintenance building (not shown) to complement the landscape design of the City Garden that had already been developed by Nelson Byrd Woltz Landscape Architects. The brief also included incorporating two pieces of sculpture: 'Femmes au perroquet' by Leger and 'Adam & Eve' by Niki De Saint Phalle, a piece that had to be protected from the environment.

The design solution was to create a glass box with an asymmetrical steel trellis to protect the glass from solar gain from the south and west, but also kept the café space visible from Chestnut Avenue (one way street heading east). Contrasted against this box is a solid stone clad box containing the kitchen and support spaces that provides a mounting location for the Leger sculpture on a major axis of the park. Because of the visibility of the buildings from the high rises surrounding the site a green roof system was incorporated.

The building is located along a raised terrace that provides outdoor dining space, accessed through a 20' long barn door on the front façade of the café, that overlooks a basin that leads to a water fall and views of the overall sculpture garden.

The dining area of the building is exposed steel frame with a steel frame window system from Hopes Windows, with flooring from the same large flamed granite pavers as the surrounding upper terrace of the garden.

The kitchen and support areas are wrapped in limestone panels installed in a rain screen system over a reinforced concrete wall system, with steel edging and trim pieces.

大门基金会委托达勒姆建筑事务所为城市花园（由尼尔森·伯德·沃兹景观建筑事务所开发设计）设计一座咖啡厅和养护大楼（未展示）。同时，设计要求融入两件雕塑作品：雷吉尔的"派罗奎特的妻子"和妮基·桑法勒的"亚当与夏娃"（一件必须受到环境保护的作品）。

设计打造了一个玻璃盒子结构，并利用不对称钢铁格栅来保护玻璃不受西南两面的太阳辐射，同时让咖啡厅在策士纳大道上（朝东的单行道）具有醒目的位置。与玻璃结构相对的是密闭的石头结构，里面设置着厨房和辅助空间，而雷吉尔的雕塑也正好悬挂在石头墙面，位于公园的中轴上。由于从周边的高层建筑上可以看到项目屋顶，设计引入了绿色屋顶结构。

建筑沿着提供露天就餐平台而建，通过一个20英寸（约0.5米）长的专门进入咖啡厅。咖啡厅远眺着通往瀑布和雕塑公园的低地。

就餐区采用钢框玻璃窗系统，地面则采用了与花园露台相同的大块火焰花岗岩铺面。

厨房和辅助区域包裹石灰石板之中，石板通过钢缘和拐角固定在钢筋混凝土结构上的雨幕系统上。

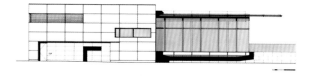
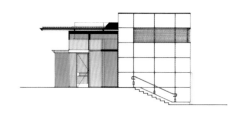
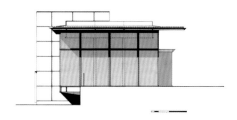

1. Night view of the sculpture in front of the café
2. Spillway in front of café
3. Southeast corner
4. The café with glass structure is eye-cathing in Chestnut Avenue.

1. 咖啡厅前的雕塑夜景
2. 咖啡厅门口的泄洪道
3. 东南角
4. 玻璃结构的咖啡厅在策士纳大道上具有醒目的位置

5. Outdoor dining terrace
6. Entrance to the café
7. Arch view down market
8. An asymmetrical steel trellis to protect the glass from solar gain from the south and west

5. 露天就餐平台
6. 咖啡厅的入口
7. 远处的大拱门
8. 利用不对称钢铁格栅来保护玻璃不受西南两面的太阳辐射

9. Café's door is open.
10. The dining area blends into the sculpture.

9. 咖啡厅的大门是敞开的
10. 咖啡厅就餐区融入雕塑作品

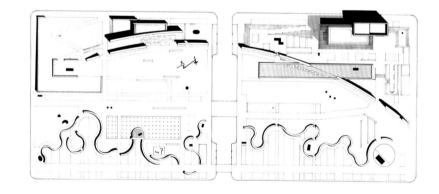

Main Level Plan
1. Entry vestibule
2. Dining room
3. Sculpture
4. Transaction counter
5. Bar
6. Back bar
7. Kitchen
8. Sculpture
9. Upper terrace
10. Sidewalk

主要层平面图
1. 入口门廊
2. 餐厅
3. 雕塑
4. 柜台
5. 吧台
6. 后部吧台
7. 厨房
8. 雕塑
9. 上层平台
10. 人行道

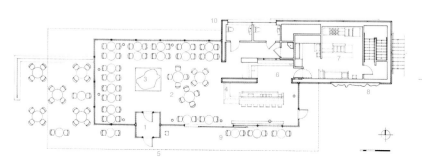

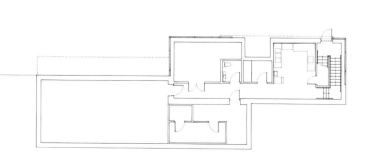

岛上大餐桌 Les Grandes Tables de L'île

Completion date: 2011 **Location:** Paris, France **Site Area:** 300 sqm **Designer:** 1024 architecture
Photographer: Brice Pelleschi & C.Sancereau

竣工时间：2011年 项目地点：法国，巴黎 占地面积：300平方米 设计师：1024建筑事务所 摄影师：布莱斯·佩莱施&C·桑瑟罗

A restaurant/bar/open-air café positioned on Île Seguin in the middle of a temporary garden whilst waiting for the architect Jean Nouvel's macro project to be implemented, Les Grandes Tables de L'ile is a place to meet for haute cuisine and even parties to accompany the reconstruction of this island steeped in history. The project is an architectural hybridization between an agricultural greenhouse, a barge and a timber-frame house. Modelled after a large wood fibre box suspended in a scaffold structure from which freight containers are hanging, all encompassed beneath a transparent umbrella... An eye-catching iconoclastic assemblage with an area of 300 sqm to accommodate120 covers and the cuisine of Arnaud Daguin, a chef with stars to his name.
Constructed from scaffolding, wood fibre panels and containers, according to the principle dear to the 1024 duo, the restaurant can be promptly extended by video and lighting effects by changing with the assistance of mapping for the duration of a party or a particular event. 'A meeting place aimed at initiating the reoccupation of the venue. An architecture which must be able to disappear without leaving any traces...'

这个集餐厅、酒吧、露天咖啡厅为一体的项目位于赛甘岛上一个临时花园（花园在等待建筑师让·努维尔的宏观项目完工）的中心，是会面、享用高档美食乃至举办派对的极佳场所。项目是农业温室、游艇和木框架住宅的结合体；悬挂在脚手架上的巨型木纤维盒子旁边悬挂着集装箱，所有结构都被一把透明伞遮挡住。作为一个占地300平方米的标新立异的项目，餐厅拥有120种菜式，而主厨阿尔诺·达根也是名声在外。
建筑由脚手架、木纤维板和集装箱组合而成，根据1024建筑事务所的设计，餐厅可以根据派对或特殊活动的需要迅速展开影音和灯光效果。"一个让场地复兴的会面场所。一座可以迅速消失、不留痕迹的建筑……"

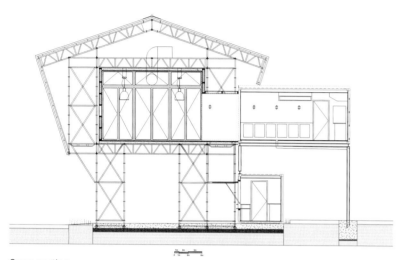

Cross section
横截面

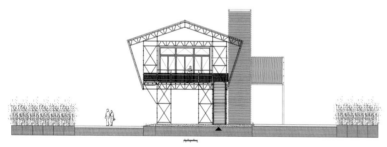

East façade
东立面

1. A night view of the architecture
2. The café is positioned on Île Seguin in the middle of a temporary garden.
3. Landscape path to the architecture

1. 建筑夜景
2. 项目位于赛甘岛上一个花园中心
3. 通向建筑的花园景观路

4

5

4. Modelled after a large wood fibre box suspended in a scaffold structure from which freight containers are hanging.
5. View the project from the garden
6. Main entrance
7.8. Connected to the scaffolding, the restaurant can be promptly extended by video and lighting effects.
9. The staircase uses the same materials as the scaffolding.

4. 悬挂在脚手架上的巨型木纤维盒子旁边悬挂着集装箱
5. 从花园看项目
6. 建筑入口
7、8. 脚手架与建筑外观相连接,餐厅可以根据需要迅速展开影音和灯光效果
9. 楼梯与建筑上的手脚架材料相同

10. The architecture is constructed from scaffolding, wood fibre panels and containers.
11. The scaffolding forms a transparent umbrella, enveloping the whole structure.

10. 建筑由脚手架、木纤维板和集装箱组合而成
11. 金属的手脚架形成透明伞，笼罩整个建筑

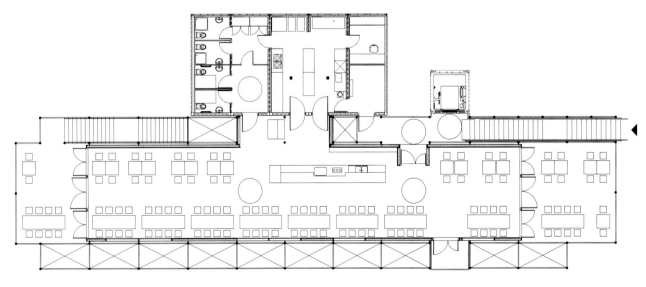

Ground floor plan
一层平面图

玛德琳餐厅 Pavilion Madeleine

Completion date: 2010 **Location:** Kayl-Tétange, Luxembourg **Area:** 220 sqm **Designer:** WW+ architektur + management **Photographer:** Linda Blatzek Photography, Trier (L/D) **Client:** Community of Kayl-Tétange, Luxembourg

竣工时间：2010年 项目地点：卢森堡，凯尔-泰坦格 项目面积：220平方米 设计师：WW建筑事务所 摄影师：琳达·布拉泽克摄影 委托人：卢森堡凯尔-泰坦格区

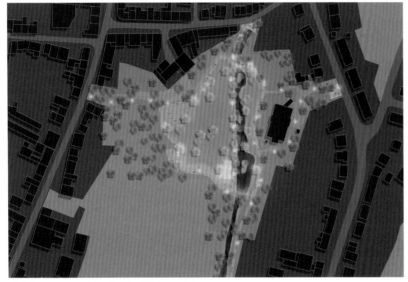

Lighting concept- Site plan by WichArchitekten

1. Pavilion Madeleine and wooden playground in the snow
2. Architecture and Landscape
3. Park Ouerbett and Pavilion Madeleine with open sun shading glass

1. 雪中的玛德琳餐厅和木操场
2. 建筑和周围景观
3. 公园和打开遮阳玻璃的建筑

The Pavilion Madeleine and the newly designed Park Ouerbett together form the new centre of the community of Kayl-Tétange, located in the south of the Grand Duchy of Luxembourg, near the industrial city of Esch-sur-Alzette.

The building is located at the intersection point of the north-south road and the loop road of the park and attracts not only the park visitors with its exterior and interior structure. Its basic dimensions of 10224.5m ensure it blends harmoniously into the park structure. The structural concept relies on a pillar-girder construction, based on an insulated bottom slab. The pillars arearranged in a 22m grid and support the 10m long IPE400 girders. All the steelwork parts are bolted together.

The transparent post-and-rail façade has been equipped with sun shading glass. A Corten-steel façade covers the volume in a modest and simple cubature, which is interrupted by several storey-high glass recesses. The existing terrace in the south featuring a barbecue area and about 40 seats also defines the entrance and leads the visitor through a glass door directly into the building. There we find a restaurant with another 40 seats. Harmonious materials as well as generous ceiling heights within the restaurant ensure visitors enjoy a high-quality dining experience. The unadorned walls contrast with the blacksteel chimney construction, the representative wine cabinet and the barmade of the same material. The three elements emphasize the clear lines of the pavilion and transport its outer sturdiness to the inside. In contrast, yet blending in with these elements, the parquet floor and the three gold-colo red suspended lamps offer a warm and comfortable atmosphere, enhanced in winter by the fire in the chimney.

The restaurant has a generous professional kitchen and toilet facilities for guests and employees. The show cooking concept was developed with the collaboration of Lea Linster, the restaurant's chef and a well-known personality beyond the borders of Luxembourg. All entrances to the pavilion as well as the inside are designed barrier-free. Minimised connection details and installations hidden in the ceiling and in the walls create neutral and silent rooms.

The energy concept of the pavilion complies with today's technical requirements. The controlled ventilation system, which exchanges the indoorair ten times per hour, ensures fresh air is distributed into the restaurant via long-range blastpipes. The elements of the kitchen equipment, such as the refrigerators, employ the latest energy-saving technology. Green roof coverings, solar cells, heat pumps and ground heat collectors are just some of the keywords that round off the sustainable master plan of this building.

餐厅和新设计的"奥尔比特公园"一起被建在卢森堡南部凯尔-泰坦格区的中心，紧邻阿尔泽特河畔埃施工业区。

建筑位于一条东西向小路和公园环形公路的交叉点上，不仅以其独特的室内外结构吸引了大量的公园游客，而且与公园结构和谐地结合在一起。建筑依赖桁架结构，以绝缘底板为基础。柱子以2米x2米的网格状排列，支撑着10米长的IPE400大梁。所有钢制构件都以螺丝连接。

透明的栏杆外立面配有遮阳玻璃。柯尔顿钢外立面以简约的外观覆盖着建筑空间，其间点缀着落地玻璃窗。南侧原有的露台配有烤肉台，可容纳40个坐席，标志着餐厅的入口，引领顾客从玻璃门进入。餐厅内部同样可容纳40人。和谐的材料和高大的房间赋予了顾客高级感。未经装饰的墙面与黑钢烟囱结构、典型的酒柜和同材质的吧台形成了鲜明对比。这三种元素凸显了餐厅的简洁线条，将外部的稳健感引入了室内。与此相反，但是同样和谐的是镶木地板和三盏金色的吊灯，它们为室内提供了温暖舒适的环境，特别是在冬天烟囱的火焰掩映之下。

餐厅拥有宽敞的专业厨房和洗手间。餐厅的主人、著名厨师李·林斯特与设计师共同开发了展示厨房的概念。餐厅的入口和内部全部采用了无障碍设计。隐藏在天花板和墙壁内的最小化连接细部和装置保证了清新宁静的室内环境。

餐厅的能源概念与现代技术要求相结合。新鲜空气通过长长的送风管引入室内，每小时能够进行10次换气。厨房设施（例如冰箱）都拥有最高的技术层级，十分节能。绿色屋顶、太阳能电池板、热泵和区域收集器仅仅是项目可持续设计技术设施的一部分，建筑拥有全套的可持续开发规划。

Awards:
2012 "European and International Property Awards" (EU/INT) – Award in the category "leisure architecture"
2012 "FIABCI Prix d'Excellence Luxemburg" (L) – Award in the category "specialized projects"
2012 "Luxembourg Green Business Awards" (L) – Nomination in the category "advisory services"
2012 "Bauhärepräis OAI" (L) – Price in the category "buildings with commercial, agricultural or industrial use"
2012 "DeutscherStahlbaupreis" (D) – Distinction in the German competition for steel construction
2011 "Concours Construction Acier" (L) – Two prizes in the categories "Non-Residential Buildings" and "Sustainability" in the Luxemburgish competition for steel construction

2012年 欧洲与国际地产大奖（欧洲/国际）——奖项类别：休闲建筑类
2012年 全球年度杰出建筑金奖（卢森堡）——奖项类别：专业项目类
2012年 卢森堡绿色商业奖（卢森堡）提名——奖项类别：咨询服务类
2012年 卢森堡建筑师与工程师协会业主奖
2012年 德国钢铁建筑奖——德国钢铁建筑卓越奖
2011年 卢森堡钢铁建筑奖——获得"非住宅类建筑"和"可持续性建筑"两个类别的奖项

4. The façade covers the volume in its modest and simple cubature, which is interrupted by several storey-high glass recesses.
5. The simple lines bring the restaurant's outer robustness inside.

4. 外立面以简约的外观覆盖着建筑空间，其间点缀着落地玻璃窗
5. 餐厅内部线条简洁，将外部的稳健感引入了室内

1. Staff changing room	1. 员工更衣室
2. Staff restroom	2. 员工洗手间
3. Beverage storage	3. 饮品仓库
4. Food storage	4. 食品仓库
5. Dish washer	5. 洗碗机
6. Beverage bar	6. 饮品吧
7. Engineering room	7. 工程室
8. Cold storage room	8. 冷库
9. Open kitchen	9. 开放式厨房
10. Disabled restroom	10. 残疾人洗手间
11. Ladies' restroom	11. 女洗手间
12. Gents' restroom	12. 男洗手间
13. Corridor	13. 走廊
14. Dining area	14. 就餐厅
15. Terrace	15. 露台
16. Loop pathway	16. 环形走道
17. Bridge	17. 桥
18. Stairs near creek	18. 溪畔阶梯
19. Kaylbach creek	19. 小溪
20. Immersible waste container	20. 可浸入式垃圾桶

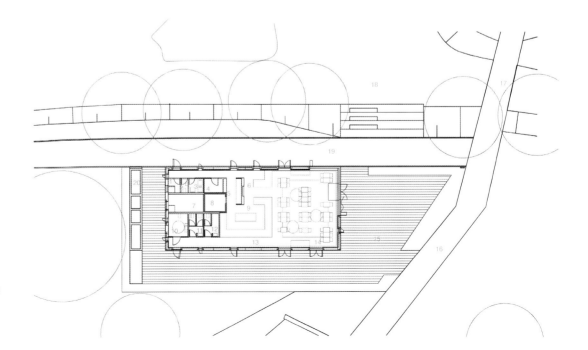

Visitor Centre / 游客中心

Guiding criteria

1. Definition

1.1 Definition. A visitor centre is a public educational facility or a dedicated space within a building for interpretive displays, programmes, services, and information. Visitor centres generally have supporting facilities and conveniences for the traveling public. Figure 1-Figure2

1.2 Interpretation. Interpretation is a combination of educational activities designed to reveal meanings and relationships through the use of presentations, original objects, firsthand experience, graphic illustrations, activities, or media designed to help people understand, appreciate, and care for the natural and cultural environment.

2. Introduction

Visitor centres are a primary type of recreation development that the Bureau of Reclamation (Reclamation) defines as a publicly recognised educational facility or dedicated space for appropriate interpretive displays and programmes. Visitor centres normally have supporting facilities (e.g., parking lots, attractive grounds, outdoor seating, walkways, and vistas) (Figure 3) and conveniences for the traveling public (e.g., toilets, water, maps, literature, telephones, and vending machines). Visitor centres (including their associated facilities, services, and programmes) serve to:
- Effectively communicate and inform the public about Reclamation and water projects.
- Enhance the quality of recreation and tourism opportunities for all visitors, including those with disabilities.
- Describe other opportunities and facilities that are available within the project.
- Provide information on the natural, cultural, and historical resources in the project area.
- Help provide visitor safety and enjoyment.
- Educate and promote water conservation and water safety.

指导标准

1. 定义

1.1 定义： 游客中心是指一个公共教育设施或在某建筑作为提供展示、规划、服务和信息的专属空间。游客中心通常有服务设施并且能为游客带来便利。图1、图2

1.2 活动方式： 游客中心是各种教育活动的结合，其宗旨是通过展示、亲身体验、图片、活动或媒体来揭示其意义和关系，帮助人们理解、欣赏自然和文化环境。

2. 简介

游客中心是娱乐开发的一个主要类型，优化改造部门将其定义为一个公认的教育设施或专用的空间，有时可适当地添加陈列品和活动。游客中心通常设有服务设施(如停车场、户外座椅、人行道等)（图3）和为游客提供的便利设施(如洗手间、饮水处、地图、游客信息导视、电话亭和自动售货机)。游客中心(包括相关设施、服务和项目)是为了：
- 有效的向游客宣传保护和节约用水的知识
- 为游客提高娱乐的质量和提供旅游平等的机会，包括残障游客
- 描述该项目其他服务机会和设施
- 提供区域的自然、文化和历史资源的信息
- 提供游客安全和愉悦的体验
- 提倡节约用水和确保游客的水上安全

Figure 1 (P72 Top)/2 (P273 Top). VanDusen Botanical Garden's new Visitor Centre is designed to create a harmonious balance between architecture and landscape, from a visual and ecological perspective. The Visitor Centre houses a café, library, volunteer facilities, garden shop, offices and flexible classroom spaces.

图1（72页 上），图2（73页 上）从视觉和生态角度来看，范杜森植物园的新游客中心在建筑和景观之间建立了和谐的平衡。通过对植物园生态的了解和分析，恢复了场地的生物多样性和生态平衡。游客中心内设有咖啡厅、图书馆、志愿者设施、花园书店、办公室和灵活的教室空间。

3. Planning

There are several tiers of guidance that may influence the planning, development, management, and operation of visitor centre facilities at Reclamation projects.

3. 规划

有几个层次的指导，可能会影响到游客中心的规划、开发、管理和运营。

Figure 3. Ticketing, drifting activities, changing rooms and washroom act cooperation with the landscape orientation and streamline arrangements.
图3 便利的服务设施、文化活动设施和展示空间规划在其平面图中。

The Resource Management Plan is the next level of guidance for planning and managing visitor centres. This plan provides comprehensive goals and objectives for project resources and can serve as a decision document to determine if a visitor centre is appropriate and suitable. A visitor centre may (or may not) be an appropriate tool to achieve the project's interpretive goals and objectives.

Interpretive master planning is the primary process for assessing whether a visitor centre is appropriate at a project. If a visitor centre is deemed appropriate, the interpretive planning process should also suggest what type of centre would be suitable. Interpretive master planning is a process that enables managers to develop a systematic and comprehensive approach to interpretation for the project or site. If a visitor centre is deemed appropriate and suitable, the interpretive master planning process can also serve to define the visitor centre messages, niche, uniqueness, accessibility, interpretive themes, interpretive tools and techniques, displays, programmes, and services. These

资源管理计划是为项目资源提供了综合的目标和宗旨，并且作确定一个游客中心是否是恰当的、合适的。

总体规划说明是评估一个游客中心是否有必要建立的基本环节。如果某个游客中心被认为是有必要建立的，总体规划中需要对游客中心的类型作出合理的建议。总体规划说明可以帮助相关管理人员为开发该游客中心，提出一个系统而全面的方案。同时，总体规划也用来规定游客中心的主题思想、位置、特色、入口的安排、景区活动的主题、展览活动和服务等。总体规划只是简单的说明规划的过程没有针对性。

Visitor Centre

guidelines briefly describe the interpretive planning process but do not address the development of specific interpretive media (e.g., exhibits, signs, or brochures).

Visitor centre proposals

1. Introduction

Constructing visitor centres is one method of providing an interpretive programme. Developing interpretive programmes and products is limited only by one's imagination. Some examples of common interpretive media are wayside exhibits, kiosks, brochures, audiotapes, videos, displays, guided hikes, nature centres, living history programmes, presentations, and visitor participatory projects.

Choosing the right interpretive medium depends on the goals of the interpretive programme, the needs of the agency, the demand and types of visitors, and the resources to be interpreted.

Visitor centres are often recommended as the desired interpretive approach before proper consideration is given to other interpretive options. In the proper environment, a visitor centre can be a very effective interpretive approach. Figure 4

However, good interpretive planning is needed to determine if and when a visitor centre should be used. Below are questions that should be answered before a decision can be made as to whether a visitor centre is the best interpretive option.

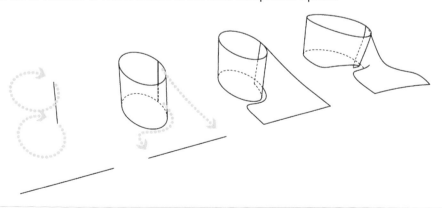

游客中心的建设

1. 简介

建设游客中心是向人们提供讲解和说明方案的一种方式。讲解和说明方案及其相关产品受限于个人的想象力。建设游客中心通常是作为很恰当的一种讲解说明的方式。如果具备合适的工作环境，游客中心可以很有效地发挥其作用。

选择正确的讲解性媒介是根据讲解性活动的目标、机构的需求、游客的类型和需求和所要讲解的资源。

在考虑选择其他适合的方法之前，游客中心是首选的讲解方式。在适当的环境中，游客中心是非常有效的讲解途径。图4

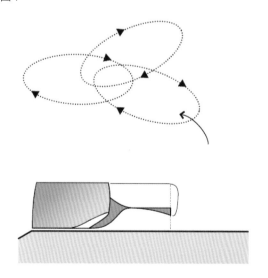

Figure 4. Thematic Pavilion EXPO 2012 Yeosu, South-Korea, Designer: soma, local partner: dmp

图4 2012世博会主题馆丽水，韩国，设计师：soma；合作伙伴：dmp

1.1 Is a visitor centre the most effective interpretive medium to use for the specific location, audience, resources, and purpose of the interpretive programme?

1.1.1 Has an interpretive plan been completed that identifies the interpretive programme goals, objectives, and themes?

- Before any design work is started on a visitor centre, there should be clear goals, objectives, and themes developed for the purpose of the visitor centre and the entire interpretive programme. A facilities planning session should be held that includes all the involved resource and accessibility specialists and potential partners. Everyone should be clear about the purpose of the overall interpretive programme and any proposed visitor centre as part of that overall programme.
- Interpretive themes should be coordinated between agencies and other local facilities so that the information is not repeated in each facility. Coordination with regional providers will help each facility complement the overall theme so the visitor has a more holistic understanding of the area.
- A visitor centre can help implement an interpretive programme. Visitor centres are very effective in providing a focus for the interpretive programmes. Tours and special events are often easier to organise when there is a visitor centre. Interpreters can use many of the visitor centre exhibits to help explain concepts and ideas before going out on the trail or taking visitors on a tour. Visitor centres, however, should not be viewed as the interpretive programme. They are only one of many strategies for reaching the public with information and education about the project and the agency's mission.
- Large visitor centre projects often result from a proposal growing beyond its original intent or from an economic development project for a local community. The success of a visitor centre as an economic development effort depends on many factors, such as proximity to major travel routes, promotional efforts, the quality of the exhibits and interpretive programme, and the potential market for the topics in the visitor centre. A market evaluation and cost-benefit analysis should be completed before design work starts on a visitor centre that is to serve as an economic development project.

转而言之，我们需要一个良好的规划来决定是否以及何时使用游客中心。下面是一些我们在选择建设游客中心作为我们的途径之前需要考虑的问题：

1.1 考虑到特定的地点，针对特定的观众，特定的资源和解释说明方案的目的，游客中心是否是最有效地一种媒介？

1.1.1 你的规划是否符合解释方案的初衷，是否体现了方案的宗旨？

- 在任何设计工作开始之前，要明确建设游客中心和这个规划设计的目的和初衷。此外，还要举行一个配套设施规划会议，会议内容包括涉及的相关资源、相关专业人士，潜在的合作者也要参与会议。每个与会人员都要知道计划的目的并认识到任何游客中心设施都是总计划的一部分。
- 所讲解的主题要在主办单位和当地的其他设施之间取得平衡，使得他们所表达的主题不会重复。与当地的供应者做好协调工作，可以对这个规划的主题起到补充完善的作用，从而使游客对该地区的历史有更好的了解。
- 游客中心可以帮助讲解活动的实施。游客中心对于讲解活动的焦点是很有效的。如有游客中心，就比较容易组织旅行。导游在外出或带游客旅行之前可以用游客中心的展品帮助讲解概念和观点。游客中心只是众多能够达到对公众传达信息和教育的战略之一。
- 大型游客中心节目往往来自于某个超越了其最初意图的提案或当地某社区的经济发展项目。一个游客中心，作为经济发展的计划，其成功取决于许多因素，如接近主要旅游路线、努力推广、展品的质量、讲解活动的质量和游客中心主题的潜在市场。对于游客中心的市场评估和成本效益分析应该在设计工作之前完成，这样才能作为一个经济开发项目。

Visitor Centre

1.1.2 Has the Interpretive Master Plan analysed the potential audiences?

- A market analysis or audience analysis regarding the types and diversity of potential visitors is critical to determining the need for a visitor centre. (See Haas and Wells, 2006.) Information may be collected using approved visitor survey instruments.
- **Demographics:** Demographics provide descriptive information about current and potential visitors (e.g., ages, genders, nationalities, incomes, disabilities, group sizes, how far they travel, as well as the social, physical, cultural, and economic factors of the area).
- **Psychographics (interests, opinions, and expectations):** Psychographics data are used to determine why visitors come to your site and what specific experiences they are seeking (e.g., being with friends and family, getting away from daily stresses, and seeking challenge in outdoor recreation).
- Type, placement, and design of a visitor centre also depend on the interests, expectations, and abilities of visitors. Visitor centres serve both orientation function and education function. Therefore, decisions about the best proportion of each function should be made during interpretive planning.
- **Orientation:** Visitor centres are very effective at orienting first-time visitors who are unfamiliar with an area and who wish to learn information about facilities, recreation opportunities, and the cultural and natural resources of an area. They are also very effective for repeat or recreational users who travel to an area for a specific recreation activity, such as rafting, fishing, or boating (Figure 5). These users typically have their equipment and are ready to start their recreation activity. Their main interest is to gain information on conditions related to their chosen recreation activity, such as where the fish are biting or the current river flows and conditions.

1.1.2 总体方案是否分析了潜在的游客?

- 对于潜在游客的类型和多样性的市场分析或游客分析对于明确游客中心的需求是至关重要的。这些主要信息来自于游客问卷。
- 人口统计:人口提供现有和潜在游客的描述性信息(如年龄、性别、国籍、收入、残障、旅行团人数、旅途距离以及这区域的社会背景、物质水平、文化和经济的因素)。
- 心理需求统计(利益、观点和期望):消费心态学的数据被用来确定游客来到游客中心的目的和他们所寻求的不寻常之旅(例如,与朋友和家人相处、远离压力、寻求户外运动的挑战)。
- 游客中心的类型、布置和设计也取决于游客的兴趣、期望和能力。游客中心要起到引导定位和教育的作用。因此,关于每项功能所占最佳比例的决策应该在规划中提出。
- 引导定位:游客中心是否能吸引首次来访的游客和那些希望了解设施信息和某区域的文化和自然资源的游客。同时,也为熟悉的游客设置特定的娱乐活动区域,比如漂流、钓鱼或划船(图5)。这些游客为了开始自己的旅途,通常带有自己的设备和做好了充足的准备。他们的主要兴趣是得到他们所选择的娱乐活动的任何信息,例如鱼在哪里比较容易上钩或者当前河道水流状况。

Figure 5. (Right) Ticketing, drifting activities, changing rooms and washroom act cooperation with the landscape orientation and streamline arrangements.
图5(右图)售票、漂流活动更衣室和卫生间呼应了景观朝向和流线安排。

Figure 6. (Left) The design aims to redefine the features of the surroundings by bringing architecture in and at the same time to dug out the potential of the surrounding background.

图6（左图）设计通过建筑的嵌入对周遭环境的特征加以重新定义，揭示出原有环境中潜藏的某些活力。

· **Education:** Visitor centres are also effective in reaching visitors with specific interests. Beyond orientation to a location, visitors are often interested in supplemental educational information about the site, its resources, or its functions. This information can include natural or cultural history. Specific goals and objectives for this level of education should be addressed in an interpretive planning process.

1.1.3 Are visitor and Reclamation goals and objectives already being served by another facility in the area?

Before a visitor centre proposal is approved, there should be a thorough survey of other visitor centres and interpretive efforts in the region. This survey should identify whether visitor needs are already met by other facilities and if Reclamation could more easily accomplish its mission by entering into a partnership with the existing facility managers. Whenever possible, visitor centres should be interagency centres. Visitors do not generally know or care about different agencies and boundaries. They usually go to a visitor centre for orientation information about an area or resource that interests them. Figure 6

1.1.4 Has the interpretive plan analysed the best location for a visitor centre?

The purpose of the interpretive programme and visitor centre is the main criterion for deciding the best location of a visitor centre. Orientation and information centres are best located at places visitors encounter before deciding where to go. If the visitor centre was built to reflect a specific regional theme, the best location will probably be near the main access. Visitors should be able to find the visitor centre easily and shortly after they enter the area. In general, poor locations for visitor centres include the end of long dirt roads, more than a few miles off the main road, deep inside the area of interest, or away from the main entrance to a resource.

1.2 Does the proposed visitor centre relate to Reclamation's mission and management objectives?

1.2.1 There should be a direct relationship between Reclamation's mission, the management objectives of Reclamation and the project, and the interpretive programme.

1.2.2 An effective visitor centre is supported by a resource management plan and is not a separate part of the overall visitor services effort. The goals and objectives of building

· **教育：** 游客中心有效地满足游客的特定利益。在某个地理位置，游客通常对选址额外补充的信息感兴趣，比如选址的资源、场地的功能。这个信息可以包括自然或文化历史。这具体目标和宗旨应该在规划过程中有所强调。

1.1.3 游客和优化改造部门的目标和宗旨是否被该地区的另一个设施所服务？

在游客中心提案审核之前，应对同区域的其他游客中心的效果进行周密调查。这调查应该鉴别其他设施是否已满足游客需求，如果和已有项目的负责人合作是否会使生态任务更容易完成。游客中心应该是尽可能跨部门合作而形成的中心。游客一般对此不感兴趣。游客经常是为了获取他们所感兴趣的信息和资源而来到游客中心的。图6

1.1.4 规划是否分析了游客中心的最佳地理位置？

建设游客中心的初衷及设置与其相关的讲解性项目的目的正是游客中心选址的主要标准。游客中心的信息导视最好位于在游客还未决定要去哪里的位置。如果游客中心的建立是为了突出某地区的主题，那么该游客中心最好的选址是主要道路的附近。在游客刚步入该地区，他们应该能够很容易地找到游客中心。一般来说，不提倡游客中心选择的位置包括：泥土路的末端、超过主道路几英里外的位置、对游客很有吸引力的区域的最里端或远离旅游资源主要入口的位置。

1.2 所提议的游客中心是否与改造任务和管理目标紧密相关？

1.2.1 改造任务、改建项目的管理目标与总体规划说明之间应该存有直接的联系。

1.2.2 一个能够发挥其作用的游客中心是由资源管理规划

Visitor Centre

a visitor centre should be clearly identified in planning documents related to the site and interpretive programme. These goals include addressing some of the challenges visitor centres bring to an area, such as the potential to concentrate or coordinate visitor use.

1.2.3 An important objective of any Reclamation visitor centre is to help the public appreciate and discover the resource diversity and recreation opportunities on Reclamation lands. In addition, helping visitors feel a sense of ownership and involvement in protecting the resources is appropriate. The visitor centre should be supported by an appropriate array of accessible current publications, exhibits, and programmes to help visitors discover and appreciate resources on nearby Reclamation lands.

2. Decision Criteria for Proposing a Visitor Centre
Arbitrary decisions are those made without principle and reason. A list of explicit decision criteria can serve several important functions in proposing and planning a visitor centre such as helping to:

· Make the decision process transparent and trackable;
· Develop a full set of reasonable alternatives;
· Ensure a full, fair, adequate, and deliberate evaluation of consequences of alternatives;
· Improve communication and increase meaningful public participation;
· Create an administrative record.

The decision criteria should fully reflect the circumstances at hand and be commensurate with the potential consequences of the decision to be made. The number of criteria needed to adequately assess development of a visitor centre increases as the potential consequences of that decision increase.

3. Interpretive Planning Process
Interpretive planning is a process of describing an existing situation and need, inventorying and analysing current resources, identifying major stories or themes, recommending a set of specific interpretive approaches and media, and implementing and evaluating products and services. It is essential for guiding the development of a visitor centre that then considers the following processes.

支持的，并不是整体游客服务中某个单独部分的成果。建筑游客中心的目标和宗旨应该清楚的体现在关于选址和规划的文件中。这些目标包括解决游客中心可能会给所在区域带来的挑战，比如游客可能过于集中，这样就涉及到协调游客如何使用设施的问题。

1.2.3 改造游客中心的一个重要目标是帮助大众欣赏并发现资源多样性和有享受休闲娱乐的机会。此外，还能使游客感到归属感并参与保护资源。游客中心应该由现有的出版物、展品和活动等一系列内容所支持，这样是为了帮助游客发现和欣赏可持续的资源。

2. 游客中心议案的决策标准
没有原则和原因的决策是武断的。一份精确的决策标准能更好地服务于规划和设计游客中心。比如：

·使决策过程透明化并且可以跟进。
·有一套完整的合理备用方案。
·确保备用方案完整和深思熟虑。
·增进交流并增加公众参与。
·创建行政记录。

决策标准应该完整地反映出目前的状况和潜在的结果。标准的数量需满足评估游客中心潜在发展的需求。

3. 规划的过程
规划是描述现有情况和需要、总结和分析当前资源、明确此规划过程的主要内容和实施策略，选出一些有针对性的方式和媒体形式，并且要对自己提供的服务进行评估和评价。此外，还应考虑一下流程：

3.1 Purpose of Planning

Why is this plan being done? This stage is often referred to as scoping and can include, but is not limited to the following.

3.1.1 Existing situation, vision, or mission of the area, resource, site, or project – What does the enabling legislation suggest about the purpose of the project? What is the mission of Reclamation in terms of resource management and visitor services? What is the existing situation that creates a demand for a visitor centre or interpretive projects or both?

3.1.2 Site or project goals – Why do interpretation at this site? Why is a visitor centre considered for this site or facility? What specific goals will the interpretation at this site or visitor centre help achieve?

3.1.3 Background or history of the area, resource, or project – What is the historic, cultural, social, and political context for planning interpretation at this place?

3.1.4 Context for planning – Are there funding, staffing, or political considerations that might influence resource management? Are there new or unusual forces exerted on the resource that necessitate interpretive planning?

3.2 Inventory and Analysis

This part of the plan inventories all the resources, issues, and audiences of the area, or park. Each of the following sections should include both an inventory of the resources and an analysis of those resources. The inventory describes what exists, and the analysis describes why that inventory is important or relevant to planning for interpretation.

3.2.1 Resource Inventory and Analysis
- Biophysical – outstanding natural and biological features.
- Socio-cultural – outstanding cultural features or phenomenon.
- Recreational resources or facilities – marinas, boat ramps, campgrounds, picnic areas, trails, etc.

3.1 规划的目的

为什么制定此规划？目的通常包括：

3.1.1 所规划的区域现有的情况如何、视野怎么样？该区域所具备的资源、地理位置在规划后能发挥什么样的作用？权威的法规关于此规划目的性能给出什么样的启示？从资源管理和游客服务的角度出发，改造任务究竟有什么？现在到底是什么催生了游客中心或讲解的活动，亦或是它同时催生了以上两者？

3.1.2 地理位置或工程的目标——为什么推行讲解计划？为什么选择此地点建立游客中心？能否帮助游客中心的运营？

3.1.3 区域、资源和工程的背景或历史——所选择地方的历史背景、文化底蕴、社会状况、政治环境是什么？

3.1.4 环境规划——资金的应用、员工的管理和政治因素是否会影响资源管理？规划中所必须的资源是否有异常？

3.2 例举和分析

规划中的这部分评估了区域所有资源和议题的状况。以下每部分都应包括现有资源的举例并对其进行分析。储备资源描述了现存资源的状况并分析了其对整体规划的重要性。

3.2.1 现有资源的举例与分析
- 生物物理方面——杰出的自然特性和生物特性
- 社会文化方面——杰出的文化标志或文化现象
- 娱乐资源或设施——游船码头、小船坡道、野营地、野餐区、山径等

Visitor Centre

3.2.2 Facilities and Programmes Inventory and Analysis
- Existing infrastructure – roads, bridges, buildings, dams, powerplants, canals, fish hatcheries, etc.
- Existing interpretation or education – interpretive exhibits and publications and/or educational collections such as skins, skulls, rocks, artifacts, and plants; library resources; and visitor orientation materials such as kiosks, bulletin boards, and orientation signs.
- Existing accommodations – provisions made for effective communication and equal opportunity to experience for persons with disabilities.

3.2.3 Management Inventory and Analysis
- Resource management summaries.
- Security issues and requirements.
- Existing plans that will affect visitor services and education.
- Any existing and relevant resource management issues (urban-wild land interface and conflicts, user conflicts between personal watercraft and anglers, sensitive natural or cultural areas, etc.) that affect the visitor experience and that should be interpreted for the visiting public.

3.2.4 Audience and Stakeholder Inventory and Analysis
- Current visitors – number of visitors, demographics, motivations, interests, market segments, etc. (NOTE: This is perhaps the most underdeveloped section of most interpretive plans.)
- Stakeholders of the area, resource, or site – partners, funders, and interest groups that might have a stake in either the management of the resources of the area or in the education of visitors to the area.

Again, it is not sufficient to just collect this information or data. Analysis involves deliberate thought, discussion, deliberation, reflection, and judgment. Consider why and how this information is useful for the project, and how this information is helpful for making decisions about this project.

3.3 Programme, Product, and Service Recommendations
This part of the plan recommends specific programmes, products, and services as they relate to the existing resource inventory, statements of significance, and visitor

3.2.2 配套设施和相关项目的列举和分析
- 现有基础设施——道路、桥梁、堤坝、发电站、运河和鱼苗孵化站等
- 目前所进行的讲解或施教的状况——陈列品和出版物；有教育意义的成品，如岩石、人工产品和植被等，图书资源，游客情况资料，如报刊亭、公告板、状况标识等
- 现有居住状况——提供良好的沟通条件并给残疾人均等的体验

3.2.3 需要进行管理的项目的列举和分析
- 总结资源管理
- 安全保障与必需品
- 现有能够影响游客接受服务和教育的规划
- 任何现有的、影响游客体验的资源管理议题（都市与荒地的衔接和冲突、私人喷气艇与钓鱼者的冲突、自然与文化的冲突等）都应有所规划

3.2.4 供求双方的状况
- 现有的游客——游客的数量、目的、兴趣和市场等。（注：这也许是规划中最被动的部分）
- 这区域的利益相关者、资源或地理位置——合伙人、基金会可能与区域资源管理或对本区域游客的教育有利害关系的利益集团

此外，不能仅仅收集信息和数据。数据要包含深度思考、讨论、审议和评估。要考虑为什么这些信息适用于此工程和怎样适用于此工程。

3.3 工程、产品和服务的推荐
规划中的这部分就现有资源分析、阐述其重要性和游客的体验而言，推荐了特定的节目、产品与服务。这建议是为了使游客中心的建设达到最佳，并且满足游客需求。同

Figure 7. 2010 Taipei International Flora Exposition, planned for different areas of the facilities for visitors

1. Rainbows
2. EXPO Dome
3. Snack seating area
4. Ticket checking
5. Ticket booth
6. Gathering plaza
7. Ticket booth
8. Yuanshan MRT station
9. Flower sea

experience opportunities. The recommendations are a strategy or prescription for the best set of programmes and services to meet the visitors' needs, while at the same time preserving the site's resource integrity. Often, the best set of programmes and services is selected from a set of recommended alternatives using criteria such as those described in the 'Decision Criteria for Proposing a Visitor Centre' section.

In an interpretive master plan, recommendations are made concerning a variety of accessible media that best meet site or park goals for visitor education. The objective is to select the most appropriate media based on available resources (time, money, personnel, and expertise) and the purposes of the plan.

3.3.1 Facilities (Figure7)
- Visitor centres.
- Kiosks.
- Waysides.
- Visitor contact stations.

3.3.2 Personal Programmes
- Guided walks and talks.
- Campfire programmes.
- Storytelling.
- Living history programmes.
- Oral histories.
- Demonstrations.
- Environmental education activities.
- Puppet shows and dramatic presentations.
- Roving interpretations.
- Visitor information stations.

3.3.3 Manufactured or Printed Products
- Publications, including Grade II Braille, audio recordings, and computer disk of text.
- Kits and adventure packs.
- Discovery boxes and travelling exhibits.

Visitor Centre

- Exhibits, including tactile features, and audio recording or computer disk of feature exhibits.
- Signs, including tactile features, and audio recording or computer disk of feature exhibits.
- Maps and brochures for self-guided activities, including Grade II Braille, audio recording, or computer disk.

3.3.4 Electronic Technology Products
- Web pages with audio description of slides provided (section 508 compliant).
- Audiotape tours with printed script.
- Video programmes (open or closed caption).
- PowerPoint slide programmes.
- High-tech programming (animatronics, augmented reality, computer interactive programmes, video-equipped microscopes, virtual reality, etc.). The section of the interpretive plan that specifies final recommendations should also include resources for the successful design, fabrication, and installation of the recommendations, including all personnel, materials and equipment, money, and time.

4. Planning
Each visitor centre must have a master plan that addresses the visitor centre facilities and programme requirements, including compliance with accessibility standards. The master plan must address each of the items listed below and be approved by the regional security officer, regional public affairs officer, and the Chief of Public Affairs.

- An inventory and analysis of current visitors and projected visitation levels;
- An inventory and analysis of existing resources to be interpreted in the visitor centre;
- The layout of the visitor centre;
- Interpretive themes and goals and a description of the method that will be used to achieve effective communication;
- Detailed recommendations for proposed interpretive exhibits and programmes (universally accessible for persons with mobility, hearing, speech, sight, or cognitive disabilities);
- A staffing plan to operate the visitor centre, taking into consideration whether and how volunteers will be used;
- Equipment needed to support exhibits and programmes;

- 提供带手稿的录音带
- (带字幕或不带字幕)的视频
- 可滑动的幻灯片演示
- 高科技程序(电子动漫、增强现实技术、计算机交互程序、微视频装备、虚拟现实等)。规划中这个部分最终推荐还应该包括成功设计的资源、制造和安装的推荐，其包括所有人员、材料和设备、金钱和时间。

4. 设计规划
每个游客中心都要有突出游客中心设施和活动需求的蓝图，这要符合参观游览标准。蓝图要强调如下的每个项目而且要经过区域安全负责人、区域公共事务官和公共事务首席官的批准。

- 现有游客和所影射的访问水平的详细报告与分析
- 游客中心对现有资源的详细报告与分析
- 游客中心的布局
- 说明布景主题和目标、对方法的具体描述，这些都将对良好地交流起到有效的作用
- 对展品和项目提出详细建议(方便所有移动、听力、语言、视力和认知上有障碍的人进入)
- 制定完善的人员配备规划，让游客中心运转。此外，还要考虑是否要用志愿者，怎样用
- 用所需的设施来支持展品和相关活动
- 经营和管理的必要预算
- 经批准所使用的费用
- 游客中心任何的合作关系
- 游客中心复审计划表
- 游客中心的安全措施与程序，包括必要的物资的和安全技术的提升

5. 旅游资源的信息管理
游客中心的运营是整个项目管理的必要和主要的部分。游

- Budget required for operation and management;
- Use of fees, if authorised;
- Any partnership supporting the visitor centre;
- Visitor centre review schedules;
- Security measures and procedures at the visitor centre, including any necessary physical and technical upgrades.

5. Visitor Centre Information

A visitor centre operation can be a necessary and integral part of total project management. The primary purpose of a visitor centre is to provide interpretive and educational information to the visiting public (including those with physical, sensory, and cognitive impairments) about the mission of Reclamation, the project and its facilities, visitor security and safety, the geographic area where the project is located, and the cultural and natural resources of the area. Visitor centres provide the necessary information for visitors to have a safe and enjoyable visit. Exhibits and other interpretive communications must be designed to stimulate interest and convey information.

The interpretive objectives of visitor centres are to:
- Enhance the public's understanding of Reclamation and its contribution to the Nation;
- Enhance the public's understanding of the history, purpose, and operation of the project and its archaeological, historical, humanmade, natural, and cultural features;
- Develop public appreciation for the proper and safe use of project resources;
- Foster the spirit of personal stewardship of public lands;
- Orient the visitor to the project and its recreational opportunities; and
- Aid project personnel in accomplishing management objectives.

Site design

1. Overall Site Design Considerations

Regardless of cost or size, contemporary visitor centre development should strive to address a number of site design concerns:

- Achieving harmony with, and ethical responsibility for, the existing surroundings, both cultural and natural. Figure 8

客中心的根本目的是为游客(包括身体、感知、认知有损害的人)提供讲解和教育的信息。这些信息包括：可持续发展的使命、相关的活动及其设施、游客的安全保证和安全性、地理位置和当地文化、自然资源。游客中心为游客提供必要的信息，使其能享有安全并快乐的旅行。展览和其他与游客交流的方式都是为了激起游客兴趣。

建设游客中心的目的是：
- 加强公众对土地再利用的理解和其对国家的重要性
- 加强公众对工程历史、目的和运营的理解，对其考古学、历史、人类制造、自然和文化的标志都有着深刻的体会
- 指导公众适当地、安全地使用工程资源
- 培养对公有土地的有个人管理工作的兴趣
- 使游客适应游客中心和其休闲的机会
- 辅助项目人员完成管理目标

规划设计

1. 规划总体设计要素

现代游客中心都应该努力强调以下因素：

- 要对现有的文化和自然环境负责，使其和谐统一。2012世博会主题馆项目选址体现其主题设计——海洋生物和海岸生活（图8）

Visitor Centre

Figure 8: The main design intent was to embody the Expo's theme The Living Ocean and Coast and transform it into a multi-layered architectural experience. The pavilion inhabits the thematic exhibition that gives visitors an introduction to the EXPO's agenda. The Best Practice Area on the upper level functions as a flexible stage for organizations and institutions. The permanent building is constructed in a former industrial harbor along a new promenade. After the EXPO the pavilion will stay an attraction for tourists and local residents.

图8 主要的设计意图是体现世博会的主题：海洋生物和海岸生活，并将它转变成多层次的建筑的体验。展馆目的是向游客介绍世博会的议程。上层是可拆卸的舞台。沿着新人行漫步道，永久建筑被构建在前工业港口。

- Maintaining both economic viability and ecological integrity throughout the entire process to the extent possible.
- Allowing simplicity of functions to prevail, while respecting basic human needs of comfort, safety, and access for persons with disabilities.
- Balancing both long- and short-term social and environmental benefits and costs.
- Minimising disturbance of cultural resources, vegetation, geology, and natural water systems.
- Identifying, as appropriate, any environmentally safe means of onsite energy production and storage in the early stages of site planning.
- Locating and orienting structures to maximise passive energy technologies.
- Providing space for processing all wastes created onsite (collection and recycling facilities, digesters, lagoons, etc.) so that reusable and recyclable resources will not be lost and hazardous or destructive wastes are considered.
- Reusing previously disturbed areas where built areas have been abandoned.
- Developing facilities to anticipate integration of energy conservation, waste reduction, recycling, and resource conservation into the visitor experience.

- 在整个工程的过程中，尽最大程度维持经济可行性和生态完整性
- 允许功能简易化，同时尊重游客对舒适和安全要求，并保证可以让残疾人进入
- 平衡短期和长期的社会和环境效益和费用
- 使游客中心对当地的文化资源、植被、地质学和自然供水系统的干扰最小化
- 在现场场地规划的早期阶段，酌情识别任何能源生产和储存且对环境安全的方式
- 定位和定向结构使被动能源技术最大化
- 为处理现场的废物提供设施和场地，（收集和回收设施、蒸炼器、潟湖等），方便资源的循环再利用，同时还应考虑那些有危险或是有破坏性的废物
- 再利用那些曾被遗弃的损害区
- 安装一系列设施，以保护能源、减少和再利用废物、保护能源

- Incorporating local materials and crafts into structures, native plants into landscaping, and local customs into programmes and operations.

2. Site Selection

Selecting a visitor centre site for Reclamation may include any of the following: reservoir, lake, beach, river, marine areas, compelling landform, scenic view, cultural resource, canal, dam, and so forth. When siting visitor facilities, consideration should be given to both natural and cultural features of an area. The site inventory and analysis should clearly identify the quality and extent of these features, possible impacts to the existing environment, and potential mitigation measures that might be necessary.

The characteristics that make an area attractive to visitors may also pose problems. Some attractive areas may be very sensitive to disturbance and unable to withstand impacts of human activity. Other attractive areas may be too remote to justify development for direct visitor use. Some areas may be too close to safety hazards or too developed to be appropriate for visitor centre development. Conversely, some degraded areas may, in fact, provide opportunities for development, allowing more options for site preservation and ecological restoration. Some areas may have terrain issues that will increase the cost of compliance with accessibility standards.

2.1 The site selection process must address the following questions:

- Will anticipated development impacts on a site be acceptable?
- What inputs (energy, materials, labour, and products) are necessary to support a development option, and are required inputs available?
- Can waste outputs (solid waste, sewage effluent, exhaust emissions) be dealt with at acceptable environmental costs?
- Will the terrain increase costs for compliance with accessibility standards (i.e., additional earthwork to meet the slope and cross slope requirements of parking spaces, accessible routes, wheelchair seating spaces in outdoor areas, required clear space at telephones, drinking fountains, waste receptacles, and other facilities)?
- The process of site selection for a visitor centre is one of identifying, weighing, and balancing the attractiveness (e.g., compelling natural and cultural features, access, and sense of place) of a site against the costs inherent in its development. The char-

游客中心

- 结构上结合当地材料和工艺，绿化上考虑本土植物，项目和操作要参考当地习俗

2. 规划

就生态而言，游客中心规划要包括以下几点：水库、湖、海滩、河、航海区域、特殊地形、风景的视野、文化资源、运河、水坝等。当为游客设施选址时，要考虑到当地的自然和文化，以及这两方面可能对现有环境造成的影响，并且提出必要的解决方案。

吸引过多游客也有可能会带来麻烦。某些吸引游客的区域也许很容易造成骚乱或者不能抵挡游客活动对其的影响。另外一些吸引游客的区域可能过于偏僻，不利于游客的直接使用。相反，那些尚未被完全开发的区域，较为完整地保存了当地的自然和生态资源，更适宜游客中心的发展。一些区域可能存在地形问题，这样就在很大程度上增加了参观的难度。

2.1 场地规划的过程必须着重考虑以下因素

- 预期的发展会影响游客中心的可访问性么？
- 对于游客中心，哪些投入是必要的（如：能量、材料、劳动和产品）？必要的投入是可行的么？
- 产生的废物(固体废物、污水、废气排放物)是可以处理成不污染环境的么？
- 为了符合易于参观游览的标准，是否会增加额外的成本？(比如在设计停车场、无障碍通道、室外轮椅座位区、电话亭、公共饮水区和垃圾桶等设施时，对纵向坡度和横向坡度有很高的要求，这就需要增加不少的土木工事)
- 为游客中心选择场地的过程，就是识别和权衡一块场地开发中的吸引人的地方(比如激发兴趣的自然环境和文化因素、入口或对场地的感觉)。传统地图或电子地图可以精确地展示地块的地理特征。那些符合环境参数的区域，可以

Visitor Centre

acteristics of a region or site should be described spatially (using either conventional or computer-generated maps) to provide a precise geographic inventory. Spatial zones meeting programmatic objectives within acceptable environmental parameters are the likely development sites.

2.2 The programmatic requirements and environmental characteristics of site development vary greatly, but the following factors should be considered in site selection.

2.2.1 Site compatibility
- visual compatibility (will the visitor centre look like it belongs in that location?)
- cultural compatibility (will the visitor centre respect local social and cultural history of the site?)
- ecological compatibility(will the visitor centre honour and/or complement the surrounding geology, vegetation, and waterforms?)

2.2.2 Visitor capacity
Every site and/or facility has a capacity for human activity. A detailed site analysis should determine this capacity based on the sensitivity of site resources, the ability of the land to regenerate, and the desired visitor experiences.

2.2.3 Density
When siting facilities, carefully weigh the relative merits of concentration versus dispersal. Natural landscape values may be easier to maintain if facilities are carefully dispersed. Conversely, concentration of structures leaves more undisturbed natural areas.

2.2.4 Climate
The characteristics of a specific climate should be considered when locating facilities so that human comfort can be maximised, while protecting the facility from climate extremes such as heat or cold, dryness, or volatile or unpredictable weather.

2.2.5 Slopes
In many environments, steep slopes predominate, requiring special sitting of structures

被用作游客中心的开发场所

2.2 选址发展的需求和环境特性变化巨大，但是下面的因素在选择场地时要考虑在内

2.2.1 选址的协调性
- 视觉的协调性(游客中心看起来像是当地建筑么?)
- 文化的协调性(游客中心是否尊重了场地的社会和文化的历史?)
- 生态的协调性(游客中心是否以周围地质、植被和水为重点或者是其补充?)

2.2.2 容纳游客的能力
每个场地或设施都有其容纳人们活动的能力。一份以地块敏感度、土地再生能力和游客的期待为基础的场地分析报告，可以评估该场地承载游客能力的大小。

2.2.3 安装设施的密度
在为场地安装设施时，应当权衡分散式安装和集中式安置的优缺点。前者由于设施分散在景区中，对自然景观影响较少，后者由于设施安装集中，只对该区域产生影响，其他区域则保持原貌。

2.2.4 气候对设施的影响
设置设施时应考虑特殊气候的特征，这样人们的舒适度就会最大化。同时，使设施远离极端气候，比如过热、过冷或干燥，还有挥发物和不可预知的天气。

2.2.5 景区环境中的斜坡因素
很多环境中，陡坡占据着很大比例，这就需要为特殊结构选址和昂贵的施工管理做考虑。在陡坡上建筑会造成水土

Figure 9: The visitor centre is a two-storey building but the lower floor is hidden by the hill, this innovation ensures that the building is well proportioned to both the hill and the reception building, and well suited to the environment.
图9 游客中心高两层，底层被山体隐藏起来，保证了建筑与山脉和建筑之间的良好比例，同时也与环境十分契合。

and costly construction practices. Building on steep slopes can lead to soil erosion, loss of hillside vegetation, inaccessible walkways and routes, damage to ecosystems, and costly ground surface impacts to provide access to persons with disabilities. Generally, appropriate site selection should locate more intense development on gentle slopes, dispersed development on moderate slopes, and no development on steep slopes.

2.2.6 Vegetation
It is important to retain as much existing native vegetation as possible to secure the integrity of the site. Natural vegetation can be an important aspect of the visitor experience and should be preserved to the degree possible.

2.2.7 Natural hazards
When considering site locations, avoid naturally hazardous situations, such as precipitous topography, animals and plants, and hazardous water areas. Site layout should allow controlled access to these features.

2.2.8 Views
Views are critical and can reinforce visitor experience. Site location should maximise desired views of natural features and desired views of facilities that support all visitor experiences.

2.2.9 Irrigation systems
Low-volume irrigation systems are appropriate in most areas as a temporary method to help restore previously disturbed areas. Irrigation piping can be reused on other restoration areas or incorporated into future domestic hydraulic systems. Captured rainwater, recycled gray water, or treated effluent should also be considered for use as irrigation water.

2.2.10 Access to natural and cultural features
Good siting practices can maximise pedestrian access to the wide variety of onsite and offsite resources and recreational activities. Low-impact development is the key to protecting vital resource areas. Figure 9

流失、山体植被减少、道路很难到达和生态系统损坏，并且代价高的地面会影响残疾人的通行。总体来说，恰当的的场地选择应该集中定位在比较缓和的斜坡上、分散在中等斜坡上、不应在陡坡上。

2.2.6 当地自然植被和游客中心的关系
尽可能多的保留当地植物有利于保护场地的完整性。在游客体验的过程中，自然植被扮演着很重要的角色，并且要尽可能的被保护。

2.2.7 自然灾害对选址的影响
当考虑选址位置时，应避开那些危险因素，比如险峻的地势、危险的动植物和有安全隐患的水域。场地的布局应该对这些因素有一定的控制作用。

2.2.8 游客中心的视野
视野是关键的并且可以增加游客阅历的。场地的位置要使自然特征和设施的视野最大化，从而增添游客的阅历。

2.2.9 灌溉系统
小体积的灌溉系统可以作为帮助恢复受损区域的临时方式。灌溉管道则被应用于恢复区域的灌溉系统或是水利循环系统中。雨水、回收的废水和已处理过的污水都可以被考虑作为灌溉水。

2.2.10 通向自然和文化的因素
好的选址可以使游客接触大量的自然和历史资源，参加各种娱乐活动。环保策略的应用对保护地块资源至关重要。
图9

Visitor Centre

2.3 Landscape considerations
Consideration of the natural landscape is important during site selection and planning. It is generally less expensive to care for landscape during construction than to restore a badly degraded landscape after construction. Placement of vegetation requires careful planning to allow growth to maturity that will not infringe on an accessible route without costly maintenance. Using native plant species and avoiding or controlling exotic or invasive species in landscape and site design are highly recommended.

3. Site Access
Site access refers to not only the means of physically entering a development, but also the enroute visitor experience. For example, the enroute experience can dramatise the transitions between origin and destination with obvious sequential gateways and can provide opportunities for interpretation or education along the way.

Other considerations for enhancing the experience of accessing a developed area include:
- Selecting corridors to limit environmental and cultural resource impacts and to control development along the corridor leading to the facility.
- Providing anticipation and drama by framing views or directing attention to landscape features along the access route.
- Providing a sense of arrival at the destination.
- Ensuring that all visitors can have the same or similar opportunities and experiences.

Site access can be achieved by various means of travel, such as by foot, private vehicles, off-highway vehicles, boats, and aircraft. Transportation means that are the least polluting, least noisy, and least intrusive in the natural environment are the most appropriate for a sustainable development. Where environmental or other constraints make physical access impossible, remote video presentation may be the only way for people to access a site.

4. Construction Methods and Materials
Construction methods and materials should be considered during the site selection process. The complexity of construction will be determined by the value of the resource, physical remoteness, and the availability of craftsmen and materials. The goal is

2.3 景观所要考虑的因素
一般来说，在建造过程中注意对自然景观的保护，比建成后恢复受损景观更节约成本。植被的布置需要仔细地规划，既要避开无障碍通道，又要易于维护。在景观和场地设计过程中，应尽量使用本地植物物种，控制和管理外来或入侵物种。

3. 规划游客中心的参观路线设计
参观游览不仅是意味着游客进入游客中心，而且还意味着游客在途中的体验。比如，旅途中可以在出发地和目的地之间戏剧化地转换并沿途提供讲解或教育的机会。

应考虑的其他因素：
- 用走廊制约对环境和文化资源的影响和掌管沿途设施
- 沿途制定视野范围或指引留意景观特征以增加游客的兴趣
- 让游客到达游客中心有抵达旅行终点的感觉
- 确保游客拥有相同的机会抵达游客中心和在其中游览有相似的体验

参观游览可以通过旅行的各种方式，比如步行、自驾游、越野车、船只和飞机。交通工具应使用最适合的可持续性发展方式，尽量做到低污染、低噪音、低干扰。当环境或其他制约使游客不愿意访问时，远程视频展示是游客进行参观游览的唯一方式。

4. 施工方法和所用材料
规划的过程中应考虑施工方法和所用材料。建设的复杂性将取决于资源的价值、距离的远近和工匠的能力以及材料的可用性。我们的目标是尽量减少对周围区域的损害，同时开发参观的设施，帮助创建一次紧密结合的、有意义的、舒适的游客体验。使用的方法和技术应该确保当项目

Figure 10. The lamellas are made of glass fiber reinforced polymers (GFRP), which combine high tensile strength with low bending stiffness, allowing for large reversible elastic deformations.
图10 薄板是由玻璃纤维增强聚合物(GFRP)组成的,其结合低抗弯刚度的高强度,允许较大可逆弹性变形。

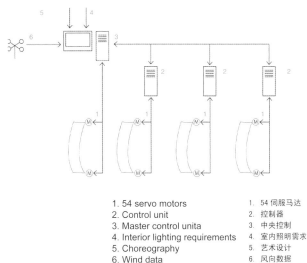

1. 54 servo motors
2. Control unit
3. Master control unita
4. Interior lighting requirements
5. Choreography
6. Wind data

1. 54 伺服马达
2. 控制器
3. 中央控制
4. 室内照明需求
5. 艺术设计
6. 风向数据

Figure 11
图11

to minimise harm to the surrounding area while at the same time, develop a visitor facility that helps create a cohesive, meaningful, and comfortable visitor experience. The methods and techniques used should ensure that there will not be unnecessary environmental damage or residual signs of construction when the project is completed. To the degree possible, the products and materials specified should be nontoxic, renewable or recyclable, and environmentally compatible with the selected site.

Primary materials are materials found in nature such as stone, earth, and flora (cotton, hemp, jute, reed, wood, and wool). If new lumber is used, consider using only lumber from certified sustainable forests or certified naturally felled trees. Use caution with any associated treatments, additives, or adhesives that may contain toxins or with materials that off-gas volatile organic compounds and thus may contribute to indoor air pollution or atmospheric pollution.

Secondary materials are materials made from recycled products such as wood, aluminium, cellulose, and plastics (Figure 10-11). Verify that production of the material does not involve high levels of energy, pollution, or waste. Verify that materials and products salvaged from old buildings are functional and safe to use. Look closely at the composition of recycled products; toxins may still be present. Consider cellulose insulation; ensure that it is fireproof and provides a greater R-value per inch thickness than fibreglass. Specify aluminium from recycled material; recycling aluminium uses 80 percent less energy to produce than initial production. Evaluate the use of products containing recycled hydrocarbon-based products; they may help keep used plastics out of landfills but may do little to reduce production and use of plastic from original resources. Keep alert for new developments; new, environmentally sound materials from recycled goods are appearing on the market every week.

Tertiary materials are manmade materials (artificial, synthetic, and nonrenewable) such as plywood, plastics, and aluminium that vary in the degrees of their environmental impact. Avoid use of materials and products containing or produced with chlorofluorocarbons or hydrochlorofluorocarbons because these chemicals deplete the ozone layer. Avoid materials that off-gas volatile organic compounds because they contribute to indoor air pollution and atmospheric pollution. Minimise use of products made from new aluminium or other materials that are resource disruptive during extraction and high energy consumers during refinement.

完成时,不会出现不必要的环境破坏或剩余工程的迹象。产品和材料应该无毒无害,可再生或可回收不破坏当地环境。

首要材料是指在自然界发现的材料,如石头、泥土和植物(棉、麻、黄麻、芦苇、木材和羊毛)。对于新木材的使用,可以考虑使用有木材认证可持续森林中的木材或已认证的自然砍伐树木。谨慎使用任何可能含有毒素的护理产品、添加剂和胶粘剂,或是排放含有挥发性和有机化合物的材料,这些材料能够造成室内空气污染和大气污染。

次要材料是由可回收产品,如木材、铝、纤维素和塑料制成的材料(图10、图11)。保证从旧建筑回收来的材料和产品是有用的和安全的。仔细查看回收产品的成分,毒素可能仍然存在。可以考虑纤维素绝缘材料,确保它是防火的,拥有比玻璃纤维更大的热阻。从回收材料中提取铝比直接生产铝,节省80%的能源。那些包含可回收碳氢化合物的产品,它们可以使塑料免于填埋,但是可能无助于降低生产和使用塑料。要时刻关注对环境无害的新型环保材料。

第三选择的材料是人造材料(人工的、合成的的、不可再生的)如胶合板、塑料、铝,它们不同程度地对环境有着影响。避免使用含有或生产氟氯化碳和氟氯化碳的材料及产品,因为这些化学物质损害臭氧层。避免使用排放挥发性有机化合物,因为它们可能造成室内空气污染和大气污染。减少使用资源破坏性材料或能源消耗高的材料。

Visitor Centre

Environmental considerations

Reclamation visitor centres may be developed in remote or urban areas. As such, the following environmental factors should be considered in building design. Management of environmental factors is important when museum property is either displayed and/or stored in a visitor centre. Visitor centre operations must comply with interior environmental management standards when museum properties are present.

1. Temperature

Temperature is a liability in climates where it is consistently too hot or too cold.

Where temperatures are predominantly too hot, building for comfortable interior temperatures may include the following suggestions:
- Maximise roof ventilation.
- Use elongated or segmented floor plans to minimise internal heat gain and maximise exposure for ventilation.
- Connect separate rooms and functions with covered breezeways.
- Maximise wall shading and induce ventilation.
- Provide shaded outdoor living areas such as porches, patios, and decks.
- Capitalise on cool nighttime temperatures, breezes, or ground temperatures.
- Avoid negative building pressurisation to reduce pounds of force required to open the door.

Where area temperatures are predominantly cold, building for comfortable interior temperatures may include the following suggestions:
- Consolidate functions into the most compact configuration.
- Insulate thoroughly to minimise heat loss.
- Minimise air infiltration with barrier sheeting, weatherstripping, sealants, and airlock entries.
- Minimise openings not oriented toward sun exposure.
- Avoid negative building pressurisation to reduce pounds of force required to open the door.

2. Sun

Direct sunshine can be a significant liability in hot climates but is rarely a liability in cold

环境因素

生态型游客中心往往位于偏远地区或城市区域。因此以下因素在设计过程中应充分考虑。游客中心规划建造需要重点考虑环境因素。游客中心建造必须符合室内环境管理标准。

1. 温度对游客中心室内建造的影响

温度与气候密切相关，当气候过热或过冷时也影响到温度的变化。

当温度太热时，要想达到舒适的室内温度可以参考以下建议：
- 使屋顶通风最大化
- 采用细长式或分段式布局不仅可以减少室内的吸热量，还能加强建筑的通风能力
- 用有篷通道连接独立式的房间
- 使墙体阴影和通风最大化
- 提供荫蔽的户外生活领域，例如门廊、庭院、甲板
- 利用凉爽的夜间温度、微风等气候条件
- 避免建筑所产生的负增压，使开门更加容易

当温度过冷时，要想达到舒适的室内温度可以参考以下建议：
- 将各个功能区尽可能紧凑的排在一起
- 彻底减少热损失
- 减少空气与障碍挡板渗透，空气分割器、密封剂、气闸条目
- 最小化开口并不是面向阳光照射
- 避免建筑所产生的负增压，使开门更加容易

2. 日照与建筑舒适度的关系

直接的阳光在炎热的气候时是比较严峻的问题，但在寒冷

climates. Sun can be an asset in cool and cold climates to provide passive heating. In either case, building design should take into account seasonal variations in solar intensity, incidence angle, cloud cover, and storm influences.

When solar gain causes conditions too hot for comfort:
- Use overhangs to shade walls and openings.
- Use site features and vegetation to provide shading to walls with eastern and western exposure.
- Use shading devices such as louvers, covered patios, and trellises with natural vines to block the sun's rays without blocking out breezes and natural light.
- Orient broad building surfaces away from the hot, late day, western sun (only northern and southern exposures are easily shaded).
- Use light-coloured wall and roofing material to reflect solar radiation (be sensitive to resulting glare and impact on natural and cultural settings).

When solar gain is to be used to offset conditions that are too cold for comfort:
- Maximise south-facing building exposure and openings.
- Increase thermal mass and envelope insulation.
- Use dark-coloured building exteriors to absorb solar radiation and promote heat gain.

3. Wind

Wind is a liability in cold climates because it strips heat away, but wind can also be a liability in hot, dry climates when it causes the human body to dehydrate and overheat. Wind can be an asset in hot, humid climates by providing natural ventilation. In designing visitor centres, use natural ventilation wherever feasible; limit airconditioning to areas requiring special humidity or temperature control such as artifact storage and computer rooms. Maximise or minimise exposure to wind through plan orientation and configuration, the number and position of wall and roof openings, and the relationship to grade and vegetation.

4. Moisture

Humidity can be a liability if, during extreme hot weather, it causes stickiness and cannot be easily evaporated away (cooling by perspiring). Strategies to reduce the discomfort of high humidity include maximising ventilation, inducing airflow around facilities, and

气候时相反。太阳可以在凉爽和寒冷气候提供被动加热。在这两种情况下，建筑设计应该考虑季节性变化中太阳能的强度、入射角、云层以及风暴的影响。

当日照过于强烈，导致温度过热而舒适度降低时：
- 使用悬臂结构，为墙体和门窗遮荫
- 充分利用场地独有特征和绿色植被，为墙体提供遮蔽
- 使用百叶窗、有盖天井与葡萄藤凉亭等遮阳设备，阻挡太阳光线，同时保证室内拥有良好的通风和充足的自然光线
- 应尽量采用南北朝向，避免东西朝向
- 采用浅色的墙体和屋顶材料，将太阳光反射回去。要注意反射过程可能产生的眩光以及对周围环境造成冲击等后果

当温度过冷，如何利用日照提高舒适度：
- 将建筑的南面墙体极其门窗最大化
- 增加热力供应和提高建筑的密闭性
- 使用深色的建筑物表面，以吸收光照和增加吸热量

3. 风与游客中心的建筑设计

风在面临寒冷气候时会产生问题，因为它吹走了热量，但风也可以在炎热干燥的气候条件下防止人体脱水和过热。风可以在湿热气候下提供自然通风。在设计游客中心时，应尽量采用自然通风。对于储藏室和机房等需要特别调整湿度和温度的空间，则应减少空调设施的数量。此外，还可以通过改变空间布局、调整墙体和门窗的位置和数量，以及利用绿色植被等方式，对受风面积进行调整。

4. 湿度与游客中心的建筑设计

在极端炎热的天气下，湿度变得尤为重要。湿度可以产生一定的黏性，不易蒸发。可以通过以下途径缓解高湿度所引发的不适：加强通风、促进空气流动，或是对厨房和浴

Visitor Centre

venting or moving moisture-producing functions, such as kitchens and shower rooms, to outdoor areas. Moisture can be an asset in hot, dry climates. Evaporation can be used to cool and humidify the air (natural air-conditioning). Techniques for evaporative cooling include placing facilities where breezes will pass over water features before reaching the facility and providing fountains, pools, and plants.

5. Storms/Hurricanes/Tornadoes

- Develop an emergency management and evacuation plan.
- Provide or make arrangements for emergency storm shelters that must also take into consideration the needs of persons with disabilities.
- Avoid development in flood plain and storm surge areas. Consider wind effects on walls and roofs.
- Provide storm shutters for openings.
- Use appropriate wind bracing and tiedowns.
- Design facilities to be safe from large storms by constructing them of light, readily available, renewable materials or design them to be constructed of sufficient mass and detail to prevent loss of life and material.

6. Rainfall

Rainfall can be a liability if concentrated runoff from developed surfaces is not managed to avoid erosion and flooding. It can be an asset if it is collected from roofs for irrigation water.

7. Topography & Seismic

Consider the building and land interface to minimise disturbance to site character, skyline, vegetation, hydrology, and soils. Consolidate functions or segment facilities to reduce the footprint of individual structures to allow sensitive placement within existing landforms. Use landforms and the arrangement of buildings to:
- Help diminish the visual impact of facilities.
- Enhance visual quality by creating a rhythm of open spaces and framed views.
- Orient visitors to building entrances.
- Accentuate key landmarks, vistas, and facilities.
- Facilitate intuitive use of the site through well-planned pedestrian access routes, which help prevent visitors from damaging areas by creating their own routes.

室等容易产生湿气的功能区进行调整。在炎热干燥的气候条件下，水分是有利的优势。蒸发作用可以用来冷却和湿润空气，可谓是天然的空调。

5. 游客中心为应对风暴/飓风/龙卷风所做出的特殊设计

- 开发应急管理和疏散计划
- 提供紧急风暴避难所，必须同时考虑到残疾人的需要
- 避免在洪泛区和风暴多发区设立游客中心，同时考虑风力对屋顶和墙体所产生的影响
- 为门窗安装防风盖
- 使用适当的抗风支撑和固定系统
- 用轻型、可再生材料设计出即使大型风暴通过时也很安全的设施，或是增加建筑的体量、加强细节处的设计，避免生命财产损失

6. 游客中心的建筑设计要考虑降雨对其影响

在没有得到有效控制的情况下，降雨可能会引发侵蚀和洪灾，成为安全隐患。

7. 游客中心的建筑设计要考虑地势与地震对其影响

减少游客中心对所在地的地理特征、植被、水文和土壤所造成的影响。对功能区和游客中心设施的位置进行调整，避免因过度分散而破坏环境。此外，合理利用当地的地形和建筑物的布局，还有诸多好处，如：
- 减少视觉冲击
- 通过创建开放的空间和观察点增强视觉质量
- 将游客引到游客中心的入口
- 突出重要的地标、景色和设施
- 按照游客的需求和习惯策划人行路，以避免游客到处乱走，对所在地造成破坏
- 确定土壤基质的类型和潜在的地震风险。使用抗震墙和适

- Determine soil substrate and potential seismic risk. Use shear walls and appropriate building anchorage and bracing details.

8. Water Bodies & Hydrology
Capture views and consider the advantages and disadvantages of off-water breezes. Minimise the visual impact of development on waterfront zones (also consider views from the water back to the shoreline). Use building setbacks and buffer zones and consider building orientation and materials. Safeguard water from pollutants, from the development itself, and from users.

Locate and design facilities to minimise erosion and impacts on natural hydrologic systems. Safeguard hydrologic systems against contamination by development and other related activities and allow precipitation to naturally recharge groundwater.

9. Pests & Wildlife
Design facilities to minimise intrusion by nuisance insects, reptiles, and rodents. Ensure that facility operators use natural means for pest control whenever possible.

Respect the importance of biodiversity. Avoid disruption of wildlife travel or nesting patterns when siting the development and try to limit construction activity as much as possible. Allow opportunities for users to observe and enjoy indigenous wildlife.

Cultural considerations

1. Archeological Resources
Preserve and interpret archeological features to provide insight into the successes and failures of previous cultural responses to the environment.

2. Local Architecture
Analyse local historic building styles, systems, and the materials used to find time-tested approaches in harmony with natural systems. Use local building material, craftsmen, and techniques to the greatest extent practicable in the development of new facilities.

3. Historic Resources
Reuse historic buildings, whenever possible, to assist in their preservation and to contribute to the special quality of the place.

当的固定和支撑系统，防止地震灾害

8．水体和水文学特征对游客中心建筑的影响
降低滨水区域游客中心所产生的视觉冲击(同时考虑其面向海岸线的视野)。避免游客中心和其使用者对水资源造成污染。

定位和设计设施来减少侵蚀和影响自然水文系统。保障水文系统不受发展及其他相关活动的污染并允许自然降水补给地下水。

9．游客中心建筑应防止虫害和保护野生动植物
采用相关设施来减少入侵的讨厌的昆虫、爬行动物和啮齿动物。确保使用自然方式来控制害虫。

尊重生物的多样性：在选址时，应避免影响野生动物的迁徙和破坏它们的食物链，同时也要减少建造活动。为游客提供去观察本土野生动物的机会。

文化因素

1．要善于利用考古遗址资源
保留和诠释了考占中所发现的文化特点，从而总结了以前人们对环境方面理解的成功与失败。

2．充分考虑地方性建筑的特色
分析当地具有历史意义的建筑形式、建筑体系和材料，创造出能与大自然和谐共存的建筑方法。要最大限度地利用当地的建筑原材料、工匠和技术，从而在新设施建筑过程中发挥实质性的作用。

3．充分利用其历史性资源
如果有可能，我们应该抽时间重新审视历史性建筑，因为这样能让我们更好地保护它们，并可以突出其所在地所具

Visitor Centre

Historic districts may be located anywhere on the installation. They often consist of significant buildings of noteworthy architecture or areas of historic significance that provide an important sense of heritage. Historic districts must also function in support of the current mission.

Preservation and enhancement of these areas are important to the overall appearance of the installation, as well as the sense of heritage and pride among military personnel. Maintaining the historical character of these areas is critical to the preservation of their visual integrity. During landscape restoration, use design solutions which complement the character of the historical style and time period.

4. Anthropology/Ethnic Background/Religion/Sociology
Understand the local culture and the need to avoid the introduction of socially unacceptable or morally offensive practices. Seek the views of the local population, as well as local, federally recognized Indian Tribes and Native American groups for design input, as well as to foster a sense of ownership and acceptance. Include local construction techniques, materials, and cultural considerations (that are environmentally sound) in the development of new facilities. Figure 12-14

5. Arts and Crafts
Incorporate local expressions of art, handiwork, detailing, and when appropriate, technology into new facility design and interior design. Provide opportunities and space for the demonstration of local crafts and performing arts.

Architectural design

Visitor centre building design considers the process of facility location, design, materials, and construction (Figure 15-16). In this process, visitor access and site entry; orientation, information, and visitor comfort needs; and programmatic needs such as educational, interpretive, and sales should all be considered. This chapter begins with an overview of general building design considerations, which is followed by guidance for visitor flow and floor planning. Finally, a series of environmental, cultural, and sensory considerations are provided as they relate to building design.

1. Overall Building Design Considerations
Once a site is selected for a visitor centre, the design of the visitor facility should:

有的特殊品质。

历史街区可能位于任何地方。它们经常包含引人注目的建筑或具有历史意义的区域，其提供了重要的传承意识。历史街区也必须服务于当下。

维护和促进这些地区对于维护游客中心的整体外观是很重要的，同时还能传承历史，增强民族荣誉感。维护这些地区的历史特征对于保护它们的视觉完整性是至关重要的。在对景观进行恢复的过程中，应使用符合其历史风格和时间特点的设计解决方案。

4. 人文/种族/宗教/社会学
充分理解当地的文化，并且要避免从事不被当地居民所接受或是冒犯当地道德规范的活动。寻求当地居民的意见、增进对于某些部落的了解，培养归属感和赞同感。在开发新的游客中心时，应采纳当地的建筑技术和材料，考虑当地的文化（应尽可能做到环保）图12-图14

5. 艺术性
如果条件具备的话，可以将当地人对艺术、手工艺、细节刻画以及技巧方面的理解和运用融入到中心设施的外观设计和内部装饰设计中。为当地工艺品和表演艺术提供展示空间和表演机会。

建筑的设计
游客中心的构造设计过程包括选址、设计、选择原材料和建造等几个方面（图15、图16）。在这过程中，应考虑如下因素：游客对游客通道和游客中心入口设置、游客中心的定位、信息量和游客对舒适度的要求；目标的需求，如教育、讲解和销售等。这章概述了一般设计考虑的因素、与建筑设计有关的一系列的环境、文化和感官因素。

游客中心

Figure 12-14. Located on the outskirts of Hangzhou, the site Liangzhu is a place surrounded by mountains and lakes, and from 3500 to 2200 BC was the centre of the Liangzhu culture. The designers' goal was to also express how it encompassed man and nature through its gentle tolerance. The floor of the Grand Chapel extends out into the solid mass of the Grand Staircase. The massive Gate Wall symbolically stands as the boundary between this sacred space and its everyday surroundings. Within the white expanse of the courtyard, subtle lines appear just by a difference in reflection. The organic form of the Curved Brick Wall interlinks with the Existing Woods as symbol of harmony between nature and people. By these colours and textures set against the natural surroundings and creating contrasting scenes, each scene can be expressed all the more clearly. The completion ceremony was a grand festival held by the local villagers.

图12-图14 良渚文化教堂坐落于杭州郊区,该地点被群山和湖泊环绕,从公元前3500年到公元前2200年都是当地文化的焦点。设计师的目标是通过其广博性来体现人与自然的和谐。教堂的地板一直延伸到室外台阶处,其宏伟的门墙极富象征意义,将教堂的神圣空间与外面的嘈杂分隔开。

- Enhance appreciation of the area (natural and cultural) and encourage or establish rules of conduct.
- Use efficient and cost-effectives technologies appropriate to the functional needs of the visitor centre (e.g., lighting, heating, cooling, waste).
- Consider cost-efficient, perhaps renewable, and compatible building materials.
- Employ a cradle-to-grave analysis in decisionmaking about construction materials and techniques (see end of chapter for more detail).
- Strive to create efficient, flexible spaces so overall building size and the resources necessary for construction and operation are minimised.
- Plan for future expansion and adaptive uses.
- Comply with all required accessibility standards for persons with disabilities.

1. 建筑设计的总体因素

一旦选定建设地点,游客中心相关设施的设计应该考虑如下因素:

- 提升区域的价值(自然和文化),确立和鼓励正确的行为准则
- 使用高效的、有成本效益的技术去迎合游客中心的各种功能(例如,照明、取暖、冷却和废物处理)
- 考虑成本效益,用可再生性和兼容性的建筑材料
- 建筑材料和技术的使用应从长远考虑(本章会对此进行详述)
- 努力创建高效的、灵活的空间,缩小建筑的规模,从而减少建造过程和运营过程所需的资源成本
- 为未来可能发生的扩建或重建指定做计划
- 遵守一切服务于残疾人的无障碍设计标准

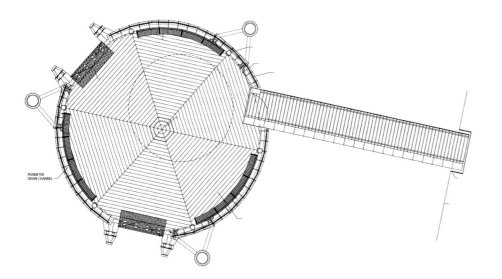

Figure 15-16. In its form, the building takes its cue from the existing semicircle of the jetty. The pavilion thus has the appearance of a seamless, circular geometrical shape which is open towards the jetty and on two sides, as well as towards the gangway pier.

图15、图16 在其形态中,建筑的灵感来自于现有的半圆形码头。展馆因此设置了无缝的、圆的几何形状外观对码头和其两边开放以及向舷梯码头开放

Visitor Centre

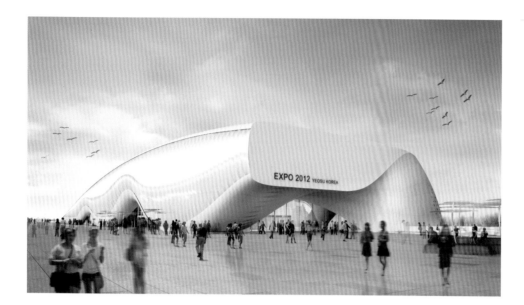

Figure 17: The main entrance is situated on Ocean Plaza, which is partly covered by the pavilion to achieve a shaded outdoor waiting area.
图17 主要入口位于海洋广场,那里部分被展馆占据,从而制造了阴凉户外等候区

2. Designing for Visitor Flow

2.1 First impressions
Visitors form initial impressions at the first encounters with the site and related facilities. Their initial reactions can influence their overall visitor experience. Gross and Zimmerman (2002) suggest the following for entry areas, parking, and walkways.

2.1.1 Entry
- Road design should follow natural contours and respect topography and landscapes. Figure 17
- Design should help slow entering vehicles and heighten awareness of surroundings.
- Road and entrance signs should be unified with those onsite, reflect the visitor centre's overall theme(s), and must comply with Reclamation's Visual Identity Programme Online Manual.

2.1.2 Parking
Parking often requires large quantities of land. Opportunities for creating people-oriented spaces are often lost around and between buildings because of expansive hard surfaced parking areas with minimum landscaping. Effective site planning and landscape design can minimise the impact of large parking areas.

The use of vegetation in parking area islands can greatly improve the visual appearance as well as help define vehicular and pedestrian circulation. Landscape islands help reduce glare and temperatures in hot climates through the use of properly spaced large shade trees. The design of parking area islands should take into consideration pavement cleaning and snow removal in northern climates. Align each island for maximum efficiency and provide sufficient area to support healthy vegetation growth.

- Parking lot placement should not impinge on the visitor centre building and should allow for transitional passage to the centre.
- A drop-off loop is often appropriate and should be provided for buses and visitors with mobility impairments.
- Service and emergency entrances and drives should be screened or routed to minimise visual impacts.
- Main parking lots should provide natural shading and landscaping that is consistent

2. 游览流程设计

2.1 第一印象
游客形成第一印象与选址和相关设施有紧密联系。他们最初的反应可以影响他们的整体游览的体验。入口、停车场和人行道等的建议如下。

2.1.1 入口
- 道路设计应遵循自然规律和尊重地形和风景。如图17
- 设计应该有助于减缓驶入车辆的速度和提高环境意识
- 道路和入口的标识应该与当地的统一,反映了游客中心的总体主题,并且必须遵守生态视觉识别在线手册

2.1.2 停车场
停车通常需要大面积的区域。建筑物之间的停车场,因为需要进行大量的硬质铺装,所以很难对其进行美化。有效的选址规划和景观设计能使大面积停车区域的影响最小化。

停车场植被的应用可以提升视觉外观并帮助区分车辆和行人的流量。通过适当间隔树荫,景观岛屿可以帮助减少太阳照射和降低温度。停车场的设计应考虑人行道的通畅性和北方气候的除雪条件。最大效率地规范并整齐排列区域,提供充足的土壤供植被茁壮成长。

- 停车场位置不应影响游客中心,而应作为通向游客中心的通道
- 应提供循环的公共汽车并为行动困难的游客提供帮助
- 服务和紧急的入口应该筛选或按路线安装在醒目的位置
- 主要停车场应提供与周围景观相协调的自然的屏障与景观
- 照明设施应覆盖游客中心周围的步行路及停车场

with landscaping throughout the rest of the site.
- Lighting should be modest; it should provide for safety but avoid light spillover. Lighting should be sufficient to light trails or walkways to and from visitor centre and parking areas.
- Accessible parking should be positioned to provide the shortest accessible route to the accessible entrance. Multiple groupings of accessible parking to serve various features are permitted.

2.1.3 Walkways
- Walkways from the parking areas to the visitor centre should be visible or clearly indicated. A view of the visitor centre is desirable.
- Walkways to the visitor centre and around the site need to consider visitor capacity, scale, and other design elements and should meet requirements under the Architectural Barriers Act of 1968 Accessibility Standards. Figure18
- A view of the visitor centre entry should be clear from major walkways. Visitors will expect to find facilities and services to meet their basic needs for information, orientation, and personal comfort. These can be provided in a number of ways.
- Each facility should meet minimum scoping and design criteria for accessibility and ensure that no services purposely or inadvertently exclude or segregate visitors in any discriminatory way.

- 停车场应位于最短的来访路径的入口。多个分组的停车服务是提倡的

2.1.3 人行道
- 从停车场到游客中心的人行道应该明显的标注。应提供平面图
- 通向游客中心及其周围的人行道设计应考虑人流量、规模和其他设计元素。如图18
- 主要人行道应清楚标明游客中心的入口。游客期望找到设施和服务，以满足他们对定位、信息和个人舒适度的基本需求。这些可以以多种方式提供
- 每个设施应满足最低范围和设计标准，并确保所提供的服务没有以任何歧视的方式有意或无意地排除或隔离游客

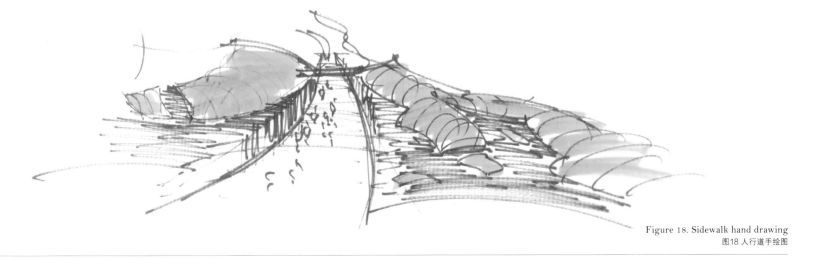

Figure 18. Sidewalk hand drawing
图18 人行道手绘图

Visitor Centre

2.1.4 Orientation and wayfinding
- After-hours information that is easy to find, well lit, and comprehensive should be provided.
- Telephones should be provided for emergency use. Public telephones should be clearly signed and meet the technical standards for persons with hearing impairments.
- Bench seating, bathrooms, and shelter in staging areas where visitors are expected to gather or wait should be provided. These staging areas should also include secure and protected areas for storing programme equipment and supplies.
- Wayfinding signs should be placed near the entrance to an area and should be on an accessible route for persons with mobility impairments. Wayfinding signs should incorporate features that aid persons with visual and cognitive impairments, such as the use of tactile characters and symbols, colour to separate and clarify themes, pictographs, and pictograms.
- Accessible features of the site should be marked with the International Symbol of Accessibility (wheelchair symbol) on the wayfinding sign.

A simple but effective sign system provides a means of communicating information without compromising the appearance of the installation. Signs are categorised as follows:

- Identification - Identifies entrance and exhibition, and other facilities.
- Destination - Directs visitors to major activities, such as the commissary, base exchange, etc.
- Regulating - Controls traffic, parking, maintains security, and identifies hazards.
- Motivational - Boosts morale, improves safety, and aids in recruiting.
- Informational - Provides educational information and directional guidance for visitors.

Installation signs are governed by AMC sign standards. Coordinate signs as a unifying landscape element of the installation's overall appearance. Vegetation should always complement and be in harmony with the sign — not overwhelm it.

2.1.5 Information Area or Lobby
- A porch or patio should be provided as an informal or formal meeting place outside the main lobby area.
- The visitor lobby should be large, open, and well lit and should provide a barrier-free

2.1.4 向导标识
- 应提供全面的非营业时间的相关信息
- 提供应急电话。公用电话应清楚的标识并且符合听力障碍人士的技术标准
- 在等待区域，应提供长椅、盥洗室、遮篷。这些区域也应该包括安全和保护区，用于存储设备和物资
- 向导标识应该放在入口处的附近区域。向导标志应当考虑有视觉和认知障碍的人，如使用盲文、颜色分开、明确的主题、图形、文字和图标
- 无障碍设施处应使用国际无障碍标志（轮椅标识）

简单有效的标识系统提供了一种交流信息的方式，而并未损害其外观。标识分类如下：

- 标志——标明入口和出口，或其他设施
- 目的地——指引游客到主要活动区域，比如商店，消费合作社
- 调节——掌控交通，停车场，维持治安并时刻警惕危险
- 激励——鼓舞士气，提高安全保障并招募义工
- 信息性——提供教育性的知识并为游客提供向导

协调标志作为安装的整体外观的统一景观要素。植被应该作为补充并且与标志和谐统——而不是将其淹没。

2.1.5 信息区与大厅

- 在主厅以外的地方应提供门廊或露台作为正式或非正式会面的地方
- 游客的大厅应该大、开放、照明充裕并应该提供一个无障碍带壁炉和地毯的入口

entry with grates and floor mats.
- Floors, walls, and ceiling surfaces should be designed to minimise noise. Different and creative floor surfaces, colours, and materials can be used to direct visitors to different areas.
- Directional signs should be large enough to be seen and should be placed where they can be seen. Use international symbols to direct visitors.
- The information desk should be brightly illuminated and barrier free (i.e., include access for wheelchairs and children).

2.1.6 Comfort Areas
- Restrooms and drinking fountains should be easy for visitors to access upon entering the visitor centre.
- Benches or appropriate seating areas should be provided around the building so visitors have several places to rest. Figure 19

- 地板、墙和天花板的表面应该隔音。不同的和创造性的地板表面、颜色和材料可用于指引游客到不同的地方
- 向导的标识应该足够大并设置在容易看到的地方。使用国际符号来引导游客
- 服务台应该是明亮的和无障碍的(即包括来访的坐轮椅的人士和儿童)

2.1.6 休息区
- 游客中心应提供卫生间和饮水机
- 建筑周围应提供长椅或适当的座位区以便游客休息。如图19

Figure 19, Benches for visitors to rest
图19 供游人休息的长椅

Visitor Centre

- Food and drink services may be considered and, if offered, should be provided in safe, comfortable, and appropriately designed areas.

2.1.7 Interpretive Media and Programme Areas
If the visitor centre is large enough to include exhibit room(s), classroom(s), or meeting room(s), consider the following:

- Exhibit room(s) should be visible and invite entry.
- Exhibits should be spaced to accommodate peak or capacity crowds.
- Exhibit space should allow for random movement rather than only directed, sequential viewing.
- Auditoriums with fixed seating are preferable for visitor centres where delivery of programmes is routine and scheduled. Multipurpose rooms with flexible seating are more appropriate for visitor centres that are used for diverse and spontaneous programming.
- Carefully consider the amount of internal space needed for circulation and how temporary seating arrangements impact occupancy loads. Maneuverability with minimum widths for accessibility is required even in temporary seating arrangements.
- The amount of space devoted to sales items should not be underestimated. Visitors value items related to their experiences. Sales and information functions, however, can often be combined for efficiency.

2.1.8 Outdoor or Onsite Areas
In almost all cases, the visitor experience extends beyond the visitor centre. Providing transition areas outdoor to enhance the visitor experiences is essential.

- Provide physical transition zones between buildings, sites, and facilities. These zones may include viewing areas, trails, interpretive waysides, or information hubs. Figure 20
- Promise adventure with outdoor site design.
- Provide outdoor activity areas and/or exhibits near the visitor centre. Create a network of opportunities or loop trails for exploring the site.
- It is important to offer the same experience and opportunities to all visitors. In the absence of accessibility guidance for trails, consider providing an accessible loop on a trail for visitors with mobility impairments that might be shorter but equally interesting. Sensory considerations and the provision of auxiliary aids are especially beneficial to

- 提供食物和饮料服务。如果提供，应提供安全、舒适、适当设计的领域

2.1.7 多媒体区域
如果游客中心足够大，包括展览的房间、教室或会议室，应考虑以下因素：

- 展览的房间应该很容易被看到并邀请大家参观
- 展品应该间隔以适应高峰的人群
- 展览的空间应该允许随意走动，而不是有针对性、有顺序地浏览
- 座位灵活的多功能厅适于进行多样化和自发性的活动。座位灵活的多功能厅适于进行多样化和自发性的活动
- 仔细考虑内部因游客流通所需的空间和临时座位安排如何影响负载。临时座位之间的安排也需要最小宽度
- 多媒体区域设置在销售区附近，游客会根据其体验对物品进行评价。使销售功能和收集信息功能相结合

2.1.8 户外区域
在几乎所有的情况下，游客的体验远远超越了游客中心。提供户外区域是提高游客的体验的至关重要的因素。

- 建筑物、设施和选址之间提供过渡地带。这些区域可能包括在观察区域、山径、路旁或信息中心。如图20
- 设计户外冒险区域
- 提供户外活动区域或在游客中心附近举办展览。提供上网服务或环山探索旅行
- 提供相同的体验和机会给所有的游客是很重要的。在缺乏无障碍指引标识的情况下，应考虑设计一种环状坡路，供残障人士使用。提供帮助给有特别需要的游客，为其制造一次同样有趣的体验

provide an equally interesting experience to visitors with visual or hearing impairments.

2.2 Accessible Design Considerations

Minimum accessibility requirements address the lowest level of access for features allowed by law and design criteria. It should be made clear that minimum requirements have a basis in law and are mandated, while Universal Design Principles are a design philosophy and are optional.

The purpose of accessibility legislation is to address human diversity and should, therefore, be incorporated into the visitor centre design to create facilities and programmes that are usable by visitors with mobility, vision, hearing, and cognitive disabilities. The standards developed to address the needs of these groups include criteria for signage, colour, text, fonts, exhibits, hierarchical text, interactive kiosks, parking, building entries, toilet rooms, slopes, cross slopes, accessible routes connecting all features, and many other considerations.

As Gross and Zimmerman (2002) suggest, **keys to achieving universal design include:**
- Integrate all visitors regardless of abilities. Do not segregate people with disabilities.

2.2 无障碍性的设计要素

无障碍设计应达到法律的最低标准，而通用设计准则可以作为参考。

无障碍设计的目的是满足人们需求的差异性，因此应该被纳入游客中心设计，从而创建可供游客(包括残疾人)使用和欣赏的设施、节目。开发的标准要包括引导标识的标准、颜色、文本、字体、展览、分级文本、报刊亭、停车场、建筑入口、洗手间、山坡、交叉山坡、易达的路线等很多其他的考虑因素。

设计的关键包括：
- 一视同仁，不要区别对待残疾人

Figure 20. The observation region of the roof
图20 屋顶的观察区域

Visitor Centre

- Provide multisensory experiences throughout the site and within the facilities to effectively communicate the information to all visitors, including those with hearing, visual, and learning impairments.
- Be flexible and creative. Consider the spirit of the accessibility laws, as well as adhering to the 'letter of the law.'
- Involve a diverse range of users and subject area specialists when designing and evaluating accessible facilities.

An excellent source for incorporating universal design into visitor centre planning and design is, Everyone's Nature: Designing Interpretation to Include All, (Hunter, 1994). The Smithsonian Guidelines for Accessible Exhibit Design (Smithsonian Institution, 1996) provide creative solutions to accessible exhibition dilemmas. The accessibility standards should be kept close at hand for reference and application to the exhibits and overall site and building design.

2.3 Sensory Considerations

Sensory considerations not only make the visit more interesting and memorable, but they will determine the success or failure of effectively communicating information to visitors (especially those with disabilities). The most effective interpretive methods employ as many of the senses as possible. Increasing the number of senses used in communication dramatically increases the effectiveness of the learning experience.

2.3.1 Visual

Provide visitors, including those with disabilities, with ready access to educational materials to enhance their understanding and appreciation of the local environment and the threats to it. Incorporate views of natural and cultural resources into even routine activities to provide opportunities for contemplation, relaxation, and appreciation. Use design principles of scale, rhythm, proportion, balance, and composition to enhance the complementary integration of facilities into the environmental context. Provide visual surprises within the design of facilities to stimulate the educational experience. Limit the height of development to preserve the visual quality of the natural and cultural landscape. Use muted colors that blend with the natural context unless environmental considerations (reflection/absorption), cultural values (customs/taboos), or safety (needed contrast for persons with visual impairment) dictate otherwise.

- 提供多种感觉体验，并使游客有效了解设施的信息，包括对那些残疾人
- 要灵活和创造。考虑法律的精神，秉承"法律条文"
- 在设计和评估设施时，要涉及不同种类的游客和专家

2.3 感官因素

设计中充分考虑感官因素的设计不仅使参观更加有趣和令人难忘，也可以确定游客(尤其是那些有残疾的人)能否进行有效交流的方式。最有效的方法是让游客有更多的感知。

2.3.1 视觉

为游客(包括那些有残疾的人)提供充足的资料，使游客更了解当地环境，避免发生潜在的危险。试图将自然和文化资源联系到日常活动中，为游客提供机会去思考、放松和欣赏。在建筑规模、设计节奏和比例以及构成因素之间的平衡等方面，运用相关的设计原则来提升设施与环境的融合度。在设计设施中提供视觉冲击以激励教育体验。限制建筑物的高度来保护自然和文化景观的视觉质量。除环境因素(反射/吸收)、文化价值观(风俗/禁忌)，或安全(需要对比视力损害的人)之外，使用柔和的颜色使其和自然环境融为一体。

2.3.2 听觉

服务和维护功能区应远离公共区域，使游客能更好的聆听自然的声音。限制使用非自然界的声音或控制其音量，比如收音机和电视。

2.3.3 触觉

允许游客触摸陈列品，从而可以直接与当地的自然与文化资源相接触。为游客(包括有视觉障碍的人)提供触觉模型，以便提供完整的游览体验。不同空间的路面应采用不同的

Figure 21 (Left). This plain/profound duality of the Ocean motivates the building's spatial and organizational concept. Continuous surfaces twist from vertical to horizontal orientation and define all significant interior spaces. The vertical cones invite the visitor to immerse into the Thematic Exhibition.

Figure 22 (Right). The open foyer is spatially defined by the twisting surfaces of the cones. The sequence of pre-show, main show and post show is spatially modulated: Lingering through the first two small cones with a ceiling height of 6 m people arrive at the main show, an breath-taking 20-metre-high space of 1000sqm.

图21（左）这海洋深远的二元性激励建筑的立体空间和组织的概念。从垂直到水平方向的连续层面转动和定义所有重要的内部空间。此垂直锥形体邀请游客沉醉于主题展览。
图22（右）开放的大厅被锥形体转动的表面所定义。展前、主秀和展后的顺序都应被空间地调制；随着前两个小锥体天花板6米高度的缓慢消失，人们到达了主秀的展馆——一间20米高、1000平方米的展馆。

2.3.2 Sound
Locate service and maintenance functions away from public areas. Space interpretive stops so that natural or site-specific sounds dominate. Use vegetation to baffle sound between public and private activities, and orient openings toward natural sounds such as the lapping of waves, babbling of streams, and rustling of leaves. Limit the use or audio level of unnatural sounds such as radios and televisions.

2.3.3 Touch
Allow visitors to touch and be in touch with the natural and cultural resources of the site. Tactile models, built to scale, offer a full experience to many visitors, including persons with visual disabilities. Vary walking surfaces to give different qualities to different spaces. Use contrasting textures to direct attention to interpretive opportunities.

2.3.4 Smell
Allow natural fragrances of vegetation to be enjoyed. Direct air exhausted from utility areas away from public areas.

2.3.5 Taste
Provide opportunities to sample local produce and cuisine.

2.4 Architectural Appearance and Interior Characteristics
Architectural appearance: The architectural appearance of visitor centres need to integrate surrounding environments as well as with eye-catching identification (Figure 21). According to the marginal theory of architecture, visitor centres have typical marginal characteristics whose architectural colours, mass, type and so on should skillfully integrate into nature environments so that they could be in harmony with natural landscape. Simultaneously, architectural form should fully reflect local cultural features and uniform regional culture situation.

Interior characteristics: Interior functions of visitor centres should be various and the flexibility of activity modes can retain visitors' interests. Therefore, interactive and static displays are favored by the visitor centres (Figure 22). Besides, rest areas, educational areas, children's areas and feedback areas could all provides visitors diverse and flexible ways of activities.

材料，并运用对比性较强的纹理加以区分。

2.3.4 嗅觉
游客在整个空间中都可以尽情地享受植物的自然芳香，通风口的设计应远离公共区域。

2.3.5 味觉
提供当地特产和菜肴。

2.4 建筑外观与内部特征
建筑外观：游客中心建筑外观除需具备醒目标识性外，还应与周围环境相协调（如图21）。边际建筑理论认为，游客中心具有典型的边际特征，其建筑色彩、体量、风格等应巧妙地融入到自然环境中，保持与自然景观的协调一致性。同时，建筑形式要充分体现本土人文特色，与地域文化氛围相融合。体量、风格等应巧妙地融入到自然环境。

内部特征：游客中心内部功能的多样化及游客活动模式的灵活性能帮助游客保持持久的兴趣。因此，交互式和静态展示都是游客中心青睐的展示方式（如图22）。另外，休息区、教育区、儿童区、意见反馈区都能为游客提供多样、灵活的活动方式。

Visitor Centre

2.5 Facilities

The construction of visitor centres should minimise the impacts of environment and the construction of functional facilities conform to the actual situation's need. Avoid duplication and redundant construction. The functional facilities of visitor centres can be divided into service facilities, management facilities, transportation facilities and infrastructure, in which service facilities are above all, including receptions, information, foods, accommodations, shoppings, entertainments, health care and other auxiliary facilities. Of course, due to different scenic spots, service facilities should be based on actual situations, such as foods or accommodations.

2.5.1 Support facilities and public use areas

Safety, visual quality, accessibility, noise, and odour are all factors that need to be considered when sitting support services and facilities. These areas need to be separated from public use and circulation areas. In certain circumstances, utilities, energy systems, and waste systems areas can be a positive part of the visitor experience. For more information, see the Utilities and Waste Systems section below.

2.5.2 Proximity of goods, services, and housing

Visitor centre developments require the input and delivery of numerous goods and services, as well as staffing for normal operation. Sitting facilities should consider the frequency, availability, and nature of these elements and the costs involved in providing them.

2.5.3 Utilities and waste systems

- **Utilities:**

With the development of a site comes the need for some level of utilities (e.g., water, waste, energy). Developments that are more elaborate have more extensive systems to provide water, waste treatment, and energy for lighting, heating, cooling, ventilating, etc. The provision of these services and the appurtenances associated with them may adversely impact the landscape and the functioning of the natural ecosystem. Early in the planning process, utility systems must be identified that will not adversely affect the environment and will work within established natural systems. After appropriate systems are selected, careful site planning and design are required to address secondary impacts such as soil disturbance and intrusion on the visual setting.

2.5 游客中心的设施

游客中心的建设应以尽量减少其对环境的影响，功能性设施的建设要符合景区的实际需求，避免重复、多余建设。游客中心的功能设施可分为服务设施、管理设施、交通设施及基础设施四大类，其中服务设施最为重要，包括接待、信息、餐饮、住宿、购物、娱乐、医疗卫生和其他辅助设施。服务设施的设置应根据不同景区的实际情况来决定，如餐饮、住宿。

2.5.1 后勤服务和公共区域

在考虑后勤服务设施的设置时，应考虑安全、质量、可访问性、噪音、气味等因素。这些区域需要与公共使用区域分离开来。在某些情况下，公用事业设备、能源系统和废物处理系统区域可以作为游客体验的一部分。

2.5.2 商品、服务和居住状况

游客中心的发展需要投入大量的人力、物力，以保证游客中心正常的运营。休息设施应考虑使用率、实用性和其自然性质，及这些因素所涉及的成本。

2.5.3 公用设备和废物处理系统

- 公用设备

公用设备是地块发展中必不可少的组成部分，它包括水处理系统、废物处理系统和能源系统（照明、加热、制冷、通风等）。提供这些服务和与它们相关的附属物可能对景观和自然生态系统的功能造成负面影响。在设计规划的早期，就应确定这些公用设备系统不会损害环境，并能在现存的自然生态系统中正常运转。选择适当的系统后，应注意在选址场地规划设计中，需要解决其造成的的二次影响，如，对土壤物理性质和电子监控系统的干扰。

▪ **Utility sitting:**
When sitting utilities, consider dispersing the facilities or using existing terrain and vegetative features to visually screen intrusive structures. In addition, aim to buffer the noise associated with mechanical equipment and the odours associated with waste treatment by manipulating the landscape through the placement of trees and shrubs. An alternative may be to feature environment-friendly utility systems for the purpose of educating the visitor.

▪ **Utility corridors:**
Because of the impacts created by utility transmission lines, onsite generation and wireless microwave receivers are preferable alternatives in many cases. When utility lines are necessary, they should be buried near other corridor areas that are already disturbed, such as roads and pedestrian paths. Where possible, locate overhead lines away from desirable view sheds and landform crests.

▪ **Plazas and courtyards:**
Outdoor pedestrian-oriented spaces are desirable and serve many functions, including the following:
· Building Entries
· Social Gathering Areas
· Recreation Areas
· Visual Appeal Areas

Inclusion of appropriate landscape elements in outdoor spaces are beneficial, whether they are an integral part of an expansive plaza or a component of a small residential courtyard. Design landscape elements to be responsive to the user and in harmony with the space. Select durable materials and always consider the maintenance and climate.

▪ **Night lighting:**
The nighttime sky can be dramatic and contribute to the visitor experience. Light intrusion and over-lighting glare can obscure night sky viewing and may disorient migratory birds. Care is required to keep night lighting to the minimum necessary for safety and security. Urban lighting standards do not apply. Low-voltage lighting with photovoltaic collectors should be considered as an energy-efficient alternative. Light

▪ 公共座椅
根据现有地形和植被特征，适当分散地设置公共座椅，使其与环境和谐统一。此外，在景观设计时，应考虑适当排列乔木和灌木，从而隔绝机械设备产生的噪音和废物处理所产生的气体。另一种选择是为了教育游客而使用环保实用系统。

▪ 输电线设施
由于输电线很容易产生视觉的冲击，因此，在很多情况下，可以选择现场发电设施或无线微波接收设器。如果必须安装输电线，则应尽量将电线埋在道路或人行道等交通处地下。此外，高架输电线的位置也应该恰当，避免阻挡游客的视野。

▪ 广场与庭院
户外的散步空间是必要的并且要带有许多服务功能区，其中包括：
· 建筑入口
· 集会区域
· 休闲区域
· 视觉展示区域

在户外空间，无论大型广场还是小型的住宅庭院，适当的景观元素是有意的。设计景观元素能满足用户需求，并与整个空间相协调。选择耐用材料并且考虑到日后的维护和气候条件。

▪ 夜间照明
夜间天空变化的多样性有助于提高游客的体验。炫目的灯光照亮了整个夜空。此外，也要达到夜间照明安全的最低标准。带有光伏收集器的低压照明系统可以作为节能的一种选择。灯具应该保持接近地面，这样可以减少眩晕的感

Visitor Centre

fixtures should remain close to the ground to minimise eye level glare. Fixtures should be of a type that directs light downward rather than outward or upward.

- **Exterior lighting**

Exterior lighting can be categorized as path, architectural, or facilities and parking lot. Through a variety of applications, lighting serves a number of functions, including the following:

· **Path Lighting** - Reinforces path hierarchy by visually differentiating major and minor path through varied light intensities, fixture types, pole spacing, and height.

· **Architectural Lighting** - Draws attention to the entrance and special features of a facility. Provides orientation and visual interest of prominent buildings or displays. Figure 23

· **Facilities and Parking Lot Lighting** - Provides safety and security, and identifies the routes and intersections. Use lighting to create a unified appearance (e.g., use light fixtures of a consistent design and lamp type to illuminate spaces surrounding a building complex). Coordinate light fixture style, scale, illumination levels, and lamp types (e.g., high and low pressure sodium, metal halide, mercury vapour, incandescent, and fluorescent) to achieve a consistent nighttime light color and unified design effect. Facilitate maintenance by selecting durable, easily accessed, and vandal-resistant fixtures.

觉。应采用能发出向下直射光的灯具，而不是向外的或向上的。

- 室外照明

室外照明包括小路、建筑、设施和停车场照明。

·小路照明——通过不同的光照强度、设备类型、杆距和高度视觉区分了主要和次要的路径，加强了小路层次结构

·建筑照明——吸引人们注意到入口和设施的特殊功能。指引杰出的建筑或展示，从而引起人们的视觉兴趣。如图23

·设施和停车场照明——提供安全保障并制定好路线。使用照明创建外观的统一（例如，设计一致的照明装置和灯具类型照亮空间围绕的综合建筑）。协调灯具样式、规模、照明度和灯具类型（如高低压钠、金属卤化物、水银蒸汽、白炽灯和荧光灯）来实现夜间灯光颜色的一致并统一设计效果。设施的维护，通过选择耐用的、容易使用的并且能抵挡蓄意破坏的固定装置

Figure 23. After sunset the analogue visual effect of the moving lamellas is intensified by linear LED bars, which are located at the inner side of the front edge of the lamella. In opened position the LED can light the neighbouring lamella depending on the opening angle.

图23 日落之后LED灯加强了移动薄板的模拟视觉效果，它位于薄板前面的内侧边缘层。在张开的位置，根据张开的角度，LED可以照亮临近薄板

- **Storm drainage:**

In a modified landscape, consideration must be given to the impacts of storm drainage on the existing drainage system and the resulting structures and systems that will be necessary to handle the new drainage pattern. The main principles in storm drainage control are to regulate runoff to provide protection from soil erosion and to avoid directing water into unmanageable channels. Removal of natural vegetation, topsoil, and natural channels that provide drainage control should be avoided to the extent practicable. An alternative is to stabilise soils, capture runoff in depressions (to help recharge groundwater supply), and revegetate areas to replicate natural drainage systems.

- **Irrigation systems:**

Low-volume irrigation systems are appropriate in most areas as a temporary method to help restore previously disturbed areas. Irrigation piping can be reused on other restoration areas or incorporated into future domestic hydraulic systems. Captured rainwater, recycled gray water, or treated effluent should also be considered for use as irrigation water.

- **Waste treatment:**

In modified landscapes, it is often appropriate to attach waste treatment systems to existing municipal systems; however, if it is not possible to attach to a municipal system, it is important to consider treatment technologies that are biological and nonmechanical and that do not involve soil leaching or major soil disturbance. While a septic system can be considered, treatment methods that result in useful products, such as fertiliser and fuels, should be investigated. Constructed biological systems are increasingly being put to use to purify wastewater. They offer the benefits of being environmentally responsive, nonpolluting, and cost effective.

2.6 Handicapped Accessibility

All areas should be barrier-free and accessible to the disabled. Consideration for the needs of the disabled is necessary for each of the previously discussed landscape site concerns. In each case, early consideration will allow for efficient and functional design and may prevent future alterations and difficult visual design problems. Handicapped accessibility is an integral part of building's entry.

- 排涝系统

在进行景观改造时，应当考虑暴雨对现有的排水系统的影响。排涝系统的主要设计原则是调节径流，从而防止水土流失，以及水流入其他渠道。排涝系统设计应该尽可能避免破坏天然植被、土壤和原有渠道。一种方式是加固土壤，减少地势低洼区域和植被区的地表径流量，以模仿天然排水系统。

- 灌溉系统

作为一个临时的方法来帮助恢复以前受损区域，小型灌溉系统是适用于大部分地区。灌溉管道可以重复使用在其他恢复区或合并到未来家用灌溉系统。雨水、再循环水或处理过的污水也应该考虑作为灌溉用水。

- 废物处理

在进行景观改造时，在现有市政设施的基础上，应设置废物处理系统。然而，如果废物处理系统不能附加在市政设施上，应考虑采用天然生物处理技术，避免造成水土流失或土壤损害。研究净化系统，使其能提炼出有用产品，如化肥、燃料。生物系统更多地应用于废水净化中，已达到保护环境、减少污染和降低成本的目标。

2.6 无障碍设计

所有区域应便于残障人士通行，每个游客中心在考虑选址时都应考虑到这一点。设计初期，就应考虑建筑的功能性，避免日后因各种问题而引起的重建。专用通道是建筑入口的不可或缺的部分。

Visitor Centre

Relations between architecture and the surrounding landscape

1. Which is primary? Architecture or Landscape?
The functions of landscape in these types of visitor centres are accentuate the architectural expression or to fill the void, the residual spaces. Landscape often provides critical connections in the urban environment. However, the role of landscape in such instances is passive and latent. In this respect, the relationship between architecture and landscape is still hierarchical.

2. Landscape is formalised?
Landscape as discourse of space and form has been attractive to architects and designers known as form-givers. The features of landscape such as continuity and flow provide new possibilities for forms and space. The implied uncertainties and openness also provide them with new design concepts and vocabularies.

3. Hidden architecture, crouching landscape
For achieving balance and harmony between landscape and architecture, one manifestation is to integrate architecture and landscape. Buildings hidden in the landscape, they are not only in form of earth-bound whose solution is environmentally rationalistic in specific locations architecture but also be driven by the symbolism of landscape.

4. Architecture in response to landscape
In establishing the mutuality between architecture and landscape, architects have chosen to focus on environment responses through architectural form, materials, and building systems themselves. Architecture and landscape don't have to be literally bonded. Reduction of energy consumption and material use, as well as utilisation of solar energy, wind, and landscape characteristics, is in itself a response to the environment and landscape.

5. Neither landscape nor architecture
Visitor centre itself is both architecture and an expression of landscape, and it is very difficult to define its bounds. This kind of area provides new landscape.

建筑和周围景观设计的关系

1. 建筑与景观的主从关系
周围景观在游客中心的设计中,衬托建筑的表现,填补剩余的空间。景观在环境中扮演着衔接性的关键功能。但景观的角色是相对被动的与潜伏的,从这一点来说,建筑与景观的关系为主从的。

2. 景观被建筑形式化
景观作为一种空间与形式的载体,对于以形式表现的建筑师而言是极具吸引力的。景观的延续的特性提示着新的空间形式的可能性,它所意含的不确定性与开放性,给建筑师一套新的概念和语义。

3. "隐藏"在景观中的建筑
寻找景观与建筑的和谐共生,其具体的体现方式就是建筑与景观融为一体。隐藏在景观中的建筑,一方面尤其在适当地点的环境合理性,另一方面也是以景观当成它的符号与形式的表现。

4. 建筑回应景观
同样是寻求一个景观与自然共生的关系,建筑师可以选择从建筑形式、材料、与营建系统本身的反思来回应自然的环境。建筑与景观不必黏在一起,降低能源与材料的使用,多利用太阳光、自然风、与景观本身的条件,减少建筑对环境的冲击,透过空间营造回到淳朴生活形态,就是对环境最好的回应。

5. 既建筑又景观
游客中心其本身既是建筑本体,也是景观的一种表现,它的界限是十分模糊的。这种灰色地带提供了新的景观。

Visitor centre and tourist safety

Tourism programmes should reflect the realities of today's threats and environment. Certain individuals and groups, if present in a Reclamation visitor centre or on a tour, have the potential to cause detriment to the facilities, employees, and other visitors. In addition, these visitors could use visitor centres and tours to gather information that might be used in a future attack or other criminal activity.

The purpose of this chapter is to provide guidelines for integrating security designs, procedures, and best practices into Reclamation visitor centres and tours to ensure the safety and security of visitors, employees, and Reclamation facilities. While these guidelines apply primarily to public tours, the information should also be used for non-public tours, as applicable.

1. Visitor Centres

1.1 Design Considerations

Building designs should reflect basic tourism security principles along with the principles of Crime Prevention through Environmental Design, also known as CPTED. These principles utilize landscaping and architectural designs to make buildings functional, aesthetically pleasing, and tourism and security friendly. (For more information on these principles, contact your Regional Security Officer or the Reclamation Security Office.) For example, rather than using highway type barriers, properly constructed and placed plant containers can be used to provide ecologically friendly stand-off distances between visitor parking areas and visitor centres.

- Plant materials should be chosen that do not provide cover for terrorists or other criminals.
- Where appropriate, monitored security cameras should be able to observe the public clearly without blockage from foliage or other visual impairments, such as blind corners or boulders.
- Windows should be architecturally pleasing, while at the same time providing protection from small flying objects, such as a thrown rock or projectile.
- Visitor centres should be designed in such a way that the public does not wait in line in areas exposed to inclement weather or potential hazardous activities.
- Public access areas should be properly illuminated. When properly implemented, lighting works both as a protective device and a method of reassurance.

游客中心和旅游安全

游客中心的总体规划应该反映当今社会存在许多潜在危害这一现实。某些出现在游客中心的特定个人或团体很可能对游客中心的设施、工作人员以及其他游客的安全带来危害。此外，这些个人或团体还可能利用游客中心收集信息，为日后犯罪活动做准备。

本章的目的是为了提供指导方针，将安全设计、游览流程和建筑设计整合起来。从而确保游客、员工和设施的安全。虽然这些方针主要应用于公共旅游，但其也同样适用于非公共旅游。

1. 游客中心

1.1 设计因素

建筑设计要体现最基本的保障旅客旅途安全的原则。这些原则利用景观和建筑方面的设计使得建筑物在功能方面、美学方面及安全方面更加人性化。例如，在建筑物中适当的安放装有植物的容器，可以给游客更加贴近自然的感觉。

- 确保种植的植物不会被犯罪分子用来隐蔽
- 尽可能确保安保设备能清晰地拍摄到公共区域，没有被树叶或其他障碍物遮挡，同时也没有拍摄死角
- 窗户的设计要体现建筑美学，但同时要阻拦飞行物体的进入，如抛出的石头
- 游客中心的设计要避免游客在排队等候时暴露在恶劣的天气或其他潜在的危险中
- 公共区域的入口处应安装良好的照明设施以起到保护的作用
- 游客中心应提供一个安静、宽敞、明亮的空间作为急救站或护士值班室

Visitor Center

- Visitor centres should include a quiet, cool, and well-lighted place that can function, if needed, as a first aid or nurse's station.
- Hallways should be designed with consideration for both rescue and evacuation needs, along with minimum accessibility requirements.
- Primary consideration should be given to how groups and individuals will be evacuated from the visitor centre and the site.
- Visitor centres should be built in such a way that non-public areas are separated as much as possible from public tourist areas. Figure 24
- All non-public areas—including employee work areas, storage rooms, or any area that is not intended for public access—should be separated from public access areas through application of a hard-line physical security strategy. Visitors should not be able to roam freely through the site's offices or facilities. Employees, contractors, volunteers, and managing partners should advise the Area Office Security Coordinator of any issues regarding visitor access to restricted areas.
- Doors leading to non-public access areas should be of solid construction, locked, access-controlled, and monitored.
- Ingress and egress areas for the staff should be separated as much as possible from those for the public.
- Restrooms and public telephones should be made available for public use within public access areas only.

- 应首要考虑游客中心的紧急疏散方式
- 应明显地划分出游客中心的非公共区域和公共区域。如图24
- 就物质安全而言，所有非公共区域，包括工作人员的工作区、储藏室或任何不对公共开放的区域都应该和公共区域划分清楚。游客不可以随意进入办公区域。员工、承包商、志愿者和或管理者应该致电该地区安全协调员办公室
- 用门锁控制非公共区域的访问并安装监控
- 为防止秩序的混乱，入口和出口要有相应的距离
- 洗手间和公用电话应设置在公共区域
- 应防止游客在旅途过程中隐藏或滞留在游客中心

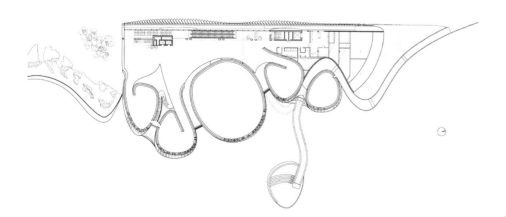

Figure 24. The entrance and exit of public area are large, while those of administration area are relatively small
图24 公共区域的出入口应比较宽阔，而行政区的出入口则相对较小

- Special consideration should be given to ensure there is no place for a visitor to hide or stay behind on a tour or within the visitor centre without being detected.

1.2 Security Measures

- As a deterrent, security measures and, where considered necessary, security forces should be readily visible to the public.
- Security forces should strive to be vigilant and engaged, but friendly and not overbearing.
- A 'No Large Bag, Backpack, or Briefcase' policy for visitors should be implemented. Small personal bags, purses, and small camera cases may be allowed on tours.
- Employees and contractors who come in regular contact with visitors should wear visible identification.
- Where access control screening has been deemed necessary, it should be performed by trained personnel, with proper access screening equipment, and supported by written procedures.
- A contingency communications system should be available. If telephone is the primary communication means, a radio, cellular phone, or similar system should be available as an alternate emergency communication system.
- Visitor centres, tour routes, and other public access areas should be periodically assessed for security-related risks. This should be accomplished as part of the formal Comprehensive Security Review and/or Periodic Security Review conducted at the host facility. However, if vulnerabilities are noted prior to the established review, they should be brought to the attention of the Area Office Security Coordinator for more immediate attention. At a minimum, the security risk assessment should address:
· Public and non-public access areas and applicable physical security measures to separate those areas.
· Tour and evacuation routes and assembly points.
· Parking areas/structures.
· Lighting and signage.
· Security and standard operating procedures for visitor management.
· Facility Security Plan coverage of visitor security.
· Integration of security procedures with the Emergency Action Plan.
· Tour guide and security officers familiarity of emergency procedures.
· Threat information applicable to the facility or general area.

1.2 安全措施

- 作为一种震慑手段，安保方面配备的设施及人员都应该安排在公众很容易发现的位置
- 安保人员应该认真负责，时刻保持警惕，要做到亲近友好而不能傲慢无礼
- 应该推行"禁止携带超长超大背包、公文包等行李"的准则；比较小的个人用背包、钱夹和相机包是可以携带的
- 员工或是业务员，如果经常为游客服务，就应该佩戴胸签，表明身份
- 特定的入口要安装适当的摄像设备进行监控，这些摄像设备要由经过专门训练的人员操作；而且，要制定相应的操作规程
- 要具备应急通讯系统。如果以电话为主要沟通手段，那么就要配备如无线电、手机或是与其相近的通讯系统，以应对突发状况
- 游客中心内部、游客的游览路线以及公共的出入口都要定期进行安全等级评估。这应该作为综合安全分析的一部分。当然，如果在分析评定之前就发现了安全隐患，应立即报告现场的安全协调人员。为降低安全风险，应遵循以下几点：
· 公共区域和非公共区域要有相应的安全隔离装置
· 设定游览路线、疏散路线和集合地点
· 设置停车场和相关设施
· 配备照明设施和路标指示
· 为了管理游客，要有安全标准和操作程序
· 有安保规划以便确保游客安全
· 安保程序要与应急安全预案统一协同执行
· 导游和安保人员要熟知紧急预案的实施程序
· 主要设施和区域内要设定安全提醒信息

Visitor Centre

1.3 Visitor Parking Areas

Tourists tend to leave valuables in their cars. When not properly managed, parking areas can become a crime magnet.

- Visitor parking should be in clearly designated parking areas only and away from places where visitors congregate or where they form lines.
- Where feasible, designated visitor parking areas should be far enough to be removed from pedestrian traffic, but close enough to be accessible to visitor centre buildings.
- Where visitors use parking areas during hours of darkness, adequate security lighting should be provided.
- Where viable, visitor-parking areas should be electronically monitored and actively patrolled by security officers.
- Evacuation routes to safe areas should be posted in parking areas.
- At a minimum, in addition to any informational signage, the following security signs should be posted and should be visible to arriving tourists:
  ・Signs that explain what items are not allowed to be brought into the visitor centre.
  ・Signs that remind tourists not to leave valuables behind, whenever possible.
  ・Signs that forbid overnight parking.
  ・Reclamation-approved crime witness signs.

2. Tours

Public tours of Reclamation facilities serve to educate and enhance the appreciation of water management while promoting the public image of Reclamation. Tours of a water or power facility are the principal means by which Reclamation can present its story to the public. Public tours are the heart of the visitor programme, but present the challenge of sharing with the public what Reclamation does and how we do it, while at the same time protecting sensitive information, Reclamation facilities, and employees.

Sustainable design considerations

Sustainable design is future oriented. The goal of sustainable buildings is to use less material, energy, and resources; produce less waste; and create healthy environments for the people who occupy them (Gross and Zimmerman, 2002). Sustainable design is important because half of the material resources used today are used in building and half of all waste production comes from construction. Some relevant sustainable considerations for sitting and planning Reclamation visitor centres include the following

1.3 停车区域

游客可能会在车内放置贵重物品，如未仔细规划，在停车场，犯罪分子会有可乘之机。

- 清楚地指定游客的停车区域。该区域应远离游客聚集的地方
- 在可行的情况下,指定的游客停车区域应该远离人行道,但要接近游客中心的建筑
- 提供停车场的照明设施
- 停车场应有监控，并派专人巡逻
- 应提供停车场安全疏散的线路图
- 至少应该让游客看到如下安全标志：
  ・禁止携带的物品
  ・提醒游客不要留贵重物品在车内
  ・禁止停车过夜
  ・注意可疑人员

2. 旅途、参观

参观不仅加强游客对水资源管理的了解，同时也改善了他们对生态和谐的理解。参观水利和电力设施可以使游客掌握其更多的信息。参观是游客中心规划的重要部分，但目前的挑战是如何与公众分享生态知识并告知应该如何去做,同时保障高机密信息、生态设施和员工的安全。

可持续设计因素

可持续的设计是面向未来的。可持续建筑的目的是使用更少的材料、能源和资源;产生更少的浪费；并为人类创建健康环境(Gross and Zimmerman, 2002)。可持续设计是至关重要，因为当今一半的物质资源应用于建筑，同时制造了大量的建筑垃圾。游客中心选址和规划中的可持续设计因素如下：

(Gross and Zimmerman, 2002):
- Consider the larger context. Is the planned facility compatible with adjacent or nearby areas?
- Build on already disturbed areas when possible; avoid sites with easily erodible soils, fragile wetland areas, or marine ecosystems; and minimise disturbance to the surrounding landscapes.
- Landscape with native materials.
- Choose sites sheltered from climatic extremes and to maximise natural cooling and heating; locate building sites to take advantage of passive solar energy.
- Use systems that channel, store, and absorb rainwater. Figure 25
- Create multipurpose access corridors for construction and eventual use by visitors and staff after the visitor centre is built.
- Use erosion barriers and tree protectors during construction.

Cradle-to-grave analysis

Sustainable design also considers building materials. The complete life cycle of resources, energy, and waste implications of possible building materials can be analysed before building construction. A cradle-to-grave analysis traces a material or product (and its byproducts) from original, raw material sources (plant, animal, or mineral) through extraction, refinement, fabrication, treatment, transportation, use, and eventual reuse or disposal. This analysis includes the tabulation of energy consumed and the environmental impacts of each action and material.

- 考虑整体环境。规划好的设施是否与周围统一
- 建立在受损区，要尽可能防止侵蚀土壤，避免在湿地地区或海洋生态地区选址，并减少对周围景观的影响
- 使用当地材料
- 选址要避免受极端气候的影响，比如过冷或过热，此时可以利用太阳能。如图25
- 在游客中心建成后，多功能通道被游客和员工使用
- 施工期间要设有栅栏来保护树木

长期目标的分析

可持续设计也应考虑建筑的材料。其资源的完整生命周期、能源和消耗，可能的建筑材料可以在施工之前进行分析。对于某种材料或产品(项目副产品)的长期分析不仅要追溯其起源，还应对其原材料来源（植物、动物或矿物）进行提取、精炼、制造、处理、再利用或处置。这个分析包括能源消耗的表格和施工与材料对环境的影响。

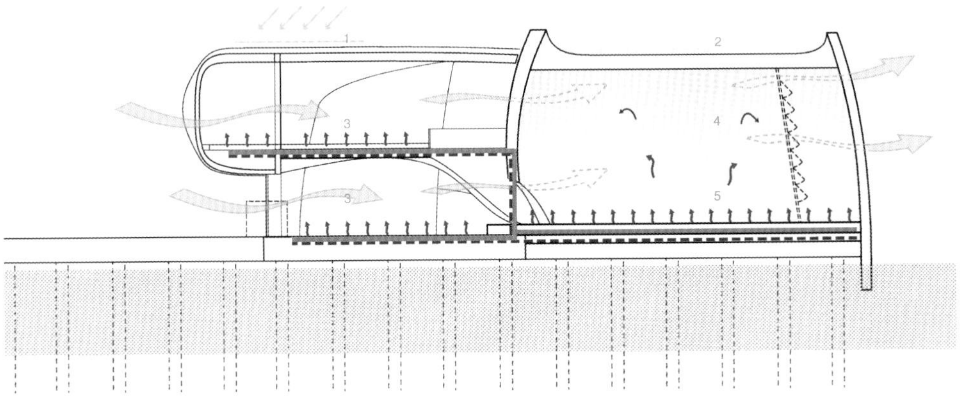

1. Solar collectors
2. Green roof
3. Natural ventilation
4. Warm air
5. Cool air

1.太阳能收集器
2.绿色屋顶
3.自然通风
4.暖空气
5.冷空气

Figure 25. Thematic Pavilion EXPO 2012 Yeosu, South-Korea. During daytime the kinetic lamellas are used to control solar input.

图25 2012丽水世博会主题馆
该项目使用了太阳能

Visitor Centre

Questions that guide a cradle-to-grave analysis include:
1. What is the source of the raw material? Is it renewable? Sustainable? Locally available? Nontoxic? Figure 26-27
2. How is the raw material extracted? What energy is used in that extraction process? What other impacts result from the extraction (e.g., habitat destruction, erosion, siltation, pollution)?
3. How is the material transported? How far does it have to be transported? How much fuel is consumed? How much air is polluted?
4. What is involved in processing and manufacturing the material? How much energy is required; what air, water, and/or noise pollution will result from the processing? What type of waste, and how much, is generated in processing and manufacture?
5. Are any treatments or additives used in the manufacture of the material? What types of treatments are necessary? Are those treatments hazardous?
6. How is the final product used? What type of energy does it require? How long will it last? How does its use affect the environment? How much waste does it generate?
7. When the product is obsolete, how is it disposed of? Can it be recycled? Does it contain solid or toxic wastes?

指导长期分析，应考虑以下问题：
1. 什么是原材料的来源？它是可再生的么？可持续吗？本地可用的吗？无毒吗？图 26、图27
2. 如何提取原材料？什么能源被用于提取过程？萃取有无其他影响(举例来说，栖息地的破坏、水土流失、泥沙淤积、污染)？
3. 如何运输材料？它的运输距离是多少？消耗多少燃料？空气污染程度如何？
4. 加工和制造材料都用到了什么？消耗多少能量？是否会污染空气和水？是否有噪音污染？产生了多少废物？
5. 是否有任何处理或添加剂用于制造材料中？什么类型的处理是必要的？那些处理危险吗？
6. 最终产品如何使用？需要什么类型的能源？它会持续多久？它的使用是如何影响环境的？产生了多少废物？
7. 当产品废弃时，如何处置？可以回收吗？它是否包含固体或有毒废物？

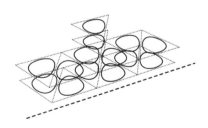 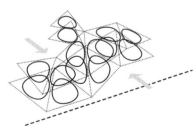 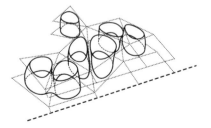 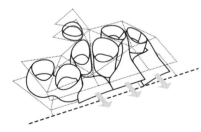

Figure 26-27: Thematic Pavilion EXPO 2012 Yeosu, South-Korea; designer:soma; local partner: dmp
The engineers of Transsolar analyzed the building's performance. Through simulations its geometry was optimised to reduce energy consumption and increase efficiency.

图26、图27 2012丽水世博会主题馆，韩国，设计师：soma；合作伙伴：dmp
Transsolar事务所的工程师们分析了建筑的性能。
通过模拟其几何形状以减少能源消耗和提高效率。建筑材料的选择应考虑当地的材料，这样可能需要更少的能源消耗来制造、运输、运作和维护材料。优先考虑材料来源可以有利于生产建筑材料的决定。

游客中心

Visitors and poisonous plants

Owing to the large amount of visitors, the toxicity of the plants aournd became extremely essential to (the) landscape and architecture. Many popular plants are considered poisonous and can produce symptoms ranging from minor to severe. This list is not exhaustive, but gives a listing of some of the most popular plantings which are known to be poisonous, as well as non-poisonous selections.

Toxic levels are based on the best information available; however, precise scientific data is not available. Toxicity is subject to numerous variables, including quantity, exposure, and individual reactions.

Plants on the high toxicity list are known to have caused death and could be hazardous with very little exposure. DO NOT USE. Plants on the medium toxicity list have toxic parts, but deaths have been rare, usually after prolonged exposure or consuming large quantities. Do not use these plants in the visitor centre. Plants on the low toxicity list include those that may cause a rash or dermititis. Use these plants with caution.

The designer shall research the toxicity of all plants specified.

游客与常见有毒植物

风景建筑,由于其游客数量众多,周围的植物的毒性变得异常重要。许多常用植物被认为是有毒的,其毒性能引起不同程度的症状。下面的列表并不全面,但提供了一些常用植物的相关信息。

毒性的级别是基于目前最新的信息;然而,精确的科学数据有时并不准确。毒性受众多不定因素的影响,包括数量、光照和个人的反应。

在高毒性列表中的植物是已知造成死亡,并且光照少也会有危险的植物。这类植物不宜使用。中度毒性列表中的植物可能含有有毒的部分,但通常经过长时间的光照能消耗掉大量毒性,很少造成死亡。游客中心中不要使用这种植物。在低毒性列表中植物包括那些可能会导致皮疹或发炎的植物。应小心使用这些植物。

设计师应研究所有植物的毒性。

HIGH TOXICITY PLANTS 高度毒性的植物

Botanical 植物科属	Common 常见植物	Toxic Part 有毒部分
Abrus Precatorius 相思树	Rosary Pea 玫瑰豆	seeds 种子
Acokanthera spectabilis/Carissa spectabilis 夹竹桃 冷杉/刺属 冷杉	Wintersweet 腊梅	fruit & plant 果实和植株
Aconitum napellus/Delphinum spp. 乌头属/飞燕草种类	Aconita, Monkshood 舟形乌头	all parts 全部
Alocasia macrorrhiza 海芋属	Cunjrvoi 海芋	all parts 全部
Brugmansia sanguinea 曼陀罗木属	Red Angles trumpet 红色喇叭花	nectar, seeds 花蜜,种子

Appendix

Botanical 植物科属	Common 常见植物	Toxic Part 有毒部分
Alocasia macrorrhiza 海芋属	Cunjrvoi 孔杰沃伊	all parts 全部
Conium maculatum 毒参	Hemlock, carrot fern or Carrot weed 铁杉木或蕨类植物或胡萝卜草	all parts, large amounts 全部，大批量
Convallaria majalis 铃兰	Lily of the Valley 幽谷百合	all parts 全部
Daphne spp. 月桂种类	Daphne 月桂	berries 浆果
Diefenbachia spp. 花月万年青属	Dumbcane 万年青	berries, few 浆果，少数
Duranta repens 杜兰丁	Duranta or Golden Dewdrop 杜兰丁或金露花	berries 浆果
Ervatamia coronaria 狗牙花	Crepe Jasmine 绉茉莉花	all parts 全部
Euphorbia pulcherrima 一品红	Poinsettia 一品红	sap 树液
Euphorbia tirucalli 绿玉书	Naked Lady or Pencil bush 藏红花	sap 树液
Gloriosa superba 嘉兰属	Glory lily 攀缘百合	all parts, esp. roots 全部，尤其是根部
Ilex spp. 冬青属	English/American Hollytree 英国、美国冬青树	fruits & leaves 果实&叶子
Jatropha spp. 麻风树属	Physic nut, Coral bush 麻风树，珊瑚灌木	seeds 种子
Kalmia spp. 山月桂属	Mountain/Western Laurel Calico Bush 山丘月桂、西方月桂 印花布灌木	all parts 全部
Laburnum anagyroides 金链花	Laburnum or Golden Chain 金链花或黄金链	all parts 全部
Lantana camara 马樱丹属	Lantana 马樱丹	green fruits 绿色果实
Lobrlia cardinalis 半边莲属	Cardinal flower 红花半边莲	all parts 全部
Lingustrum spp. 女贞	Privet 雏花	fruit 果实
Malus spp. 罗盘属	Apple 苹果树	leaves, seed in large amnt. 叶子，种子
Melia azedarch 栋树	Cape lilac or White cedar 淡紫色和白色雪松	fruit, leaves, bark, flowers 果实，叶子，树皮，花朵
Melianthus comosus 蜜花凤梨属	Tufted honeyflower 缨球花蜜	entire plant, esp. roots 整棵植物，特别是根部
Nerium oleander 夹竹桃	Oleander 夹竹桃	all parts 全部
Nicotiana glauca 烟草	Tree tobacco 树状烟草	entire plant, esp. leaves 整棵植物，特别是叶子
Prunus armeniaca 杏树	Apricot 杏	kernel in large amounts 核
Prunus dulcis 李子属	Prunus 李子	kernel-bitter type 苦核

Botanical 植物科属	Common 常见植物	Toxic Part 有毒部分
Prunus persica 桃属	Peach 桃	kernel, flower, leaf, bark 核，花，叶子，树皮
Rheum Rhaponticum 使用大黄属	Rhubarb 大黄	leaf blade 叶片
Rhododendrom 杜鹃花属	Rhododendron or Azalea 杜鹃花	leaf 叶
Ricinus communis 普通蓖麻属	Castor Oil plant 蓖麻	seeds: 2-8 种子：2~8
Solanum nigrum 龙葵	Black nightshade or Blackberry nightshade 龙葵或黑莓茄	green fruit 绿色果实
Solanum psedocapsicum 茄属植物	Madeira winter cherry or Jerusalem cherry 大西洋群岛的冬樱桃或冬珊瑚	berries 浆果
Solanum tuberosum 马铃薯	Potatoes 马铃薯	green skin 绿色表皮
Taxus baccata 紫杉属	Yew 紫杉木	all parts, esp. seed in pod 全部，特别是豆荚里的种子
Thevetia peruviana 黄花夹竹桃属	Yellow oleander 黄色夹竹桃	all parts, esp seed in kernel 全部，特别是核中的种子

NON-TOXIC PLANTS　无毒植物

No evidence currently exists that these plants are poisonous. 目前没有证明以下植物有毒

African violet. 非洲紫罗兰	Lilac 丁香	Christmas cactus 仙人掌
Marigold 金盏花	Coleus 薄荷科植物	Norfolk pine tree 松树
Corn plant 谷物	Peperomia 豆瓣绿	Crocus (spring) 番红花
Petunia 矮牵牛花	Dandelion 蒲公英	Prayer plant 巴西条纹竹芋
Dogwood 水木	Pyracantha/Firethorn 火棘	Dracaena 龙血树
Rose 玫瑰	Easter lily 复活节百合	Rubber tree plant 橡胶树
Ferns 蕨类植物	Sansevieria/Snake plant 虎尾兰	Ficus* 无花果
Scheffiera* 鹅掌柴属	Forsythia 连翘属植物	Spider plant 吊兰
Fuchsia 灯笼海棠	Swedish Ivy 瑞典常春藤	Geranium 天竺葵
Tulip* 郁金香	Hibiscus 芙蓉	Wandering Jew 白花紫露草
Honeysuckle 金银花	Wax plant 毬兰	Impatiens 凤仙花属植物
Wild strawberry/Snakeberry 泡草莓/泻根属的果实	Jade plant 青锁龙	Zebra plant 种子银脉金叶木

*Sap may be irritating
带*的树液也许会引起刺激性的作用

Pavilions in Lake Yangcheng Park, Kunshan

昆山阳澄湖公园景观建筑

Completion date: 2010 **Location:** Kunshan, China **Construction areas:** 160 sqm **Designer:** Pu Miao and Hanjia Design Group, Shanghai Yuangui Structural Design Inc. **Photographer:** Pu Miao, Other disciplines: Zhuang Wei, Xu Man, Zhou Leyan, and Li Wen

竣工时间：2010年 项目地点：中国，昆山 建筑面积：160平方米 设计师：缪朴及汉嘉设计集团，上海源规建筑结构设计事务所 建筑：缪朴，蒋宁清 摄影师：缪朴 其他工种：张业巍，刘潇，郭忠，于洋及吴秋燕

Located in a new suburban park, this is the constructed one of three proposed pavilions that share one prototypal form and structural system.

The design explored the theme 'multi-valency' in two aspects. First, Chinese traditional garden buildings took the form of courtyard which combines both interior and exterior spaces, providing users two kinds of enjoyment within a short distance. This 36X14X3.34-metre building volume contains indoor, outdoor and semi-outdoor spaces. The 14-metre span of the structure allows the designers to juxtapose these spaces freely, rather than grouping them into two chucks.

Next, double-skin glass facades line up the south and north sides of the building. The use of such a facade had been limited to high-rise office buildings and mainly for energy saving. The design tried to add two new meanings into it. The designers added two tiers of shelves in the cavity to allow potted plants to grow there, making the facade a 'green' wall (that can also be used for displays). They also partially sand blasted the glass skins to create small transparent 'windows' that overlap each other in the two skins. People will see three kinds of depths, with varied overlapping effects when the viewer moves.

Both skins of the wall are openable. In mild climates windows on both skins can be opened to afford natural ventilation. While all windows are closed in winter, only the windows in the outer skin are opened in the hot season. The shelves only cover two thirds of the cavity space to make room for the vertical air flow.

本工程是昆山市郊一个新建大型公园中的三个服务设施中先建成的一个。三个建筑将共用一个形式母题及结构体系。

本设计在两个层面上探讨了"复合"这一主题。首先，我国的传统园林建筑大多为包含室内外空间的庭院建筑，能在一个小范围内让人同时享受到不同环境各自的优点。为此，我们在这个36米×14米×3.34米高的建筑形体中复合了室内、室外及半室外空间。跨度14米的结构使我们得以在三个单体中对室内外空间按使用做交错的布置，而不再捏合成两组。

其次，建筑的南、北两面各是一道双重玻璃外墙。该类外墙到目前为止基本上局限于高层办公楼中，只为节约能源而用。我们把它与两种新的意义复合起来。在双重外墙中设计了二层搁板，使其成为可生长盆载植物的"绿色"外墙（也可陈列商品）。我们还在内外玻璃面上用局部磨砂创造出内外错开的透明"窗口"。使外墙呈现出三种层次的通透感。当人移动时，这些窗口会彼此掩映。

复合外墙的内外层均可开启。在气候宜人时可打开内外窗利用自然通风。在冬季可同时关闭内外层。在炎夏可只开启外窗。为了保证空气的垂直流动，搁板只覆盖了三分之二的空腔平面。

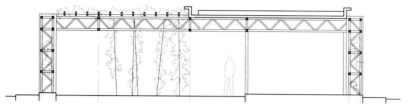

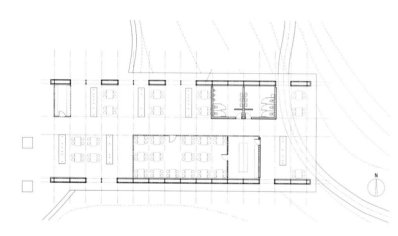

1. The three kinds of depths of the double-skin wall
2. The double-skin wall, seen from the interior
3. Northeast view of the building
4. The outdoor public resting space in the building. Rental is at the right.

1. 从室外看双重外墙的三种通透层次
2. 从室内看双重外墙
3. 从东北角望建筑全景
4. 从东南角望建筑内的室外休息空间，右面是出租空间

5. The east end of the building is penetrated by a park path.
6. The west end of the building opens into the ellipse square, the Snack Shop is to the left.
7. South elevation, part of the windows are open
8. The outdoor public Resting space in the building, seen from the north.

5. 建筑东端被现有园路穿过
6. 建筑西端开向椭圆广场,左边为小卖部
7. 建筑南立面,部分窗被开启
8. 从北望建筑内的室外休息空间,左面为出租空间

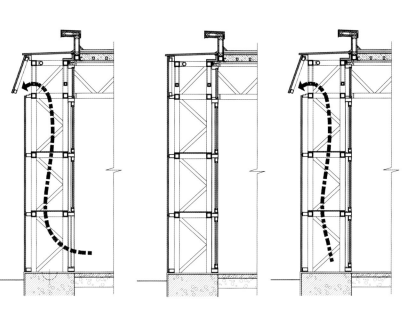

麦克道尔·索诺兰保护区入口

GATEWAY to the McDowell Sonoran Preserve

Completion date: 2009 **Location:** Arizona, USA **Construction Area:** 560 sqm **Site area:** 155 ha
Designer: Weddle Gilmore **Photographer:** Bill Timmerman, Chris Brown

竣工时间：2009年 项目地点：美国，亚利桑那州 建筑面积：560平方米 占地面积：155公顷 设计师：韦德尔·吉尔默 摄影师：比尔·蒂莫曼、克里斯·布朗

The Gateway was designed to celebrate the entry and passage into the 36,400 acre McDowell Sonoran Preserve while minimising the impact on the native desert. The Gateway is the point of access to over 45 miles of trails within the McDowell Sonoran Preserve for hiking, bicycling, and equestrian enjoyment. The project site design achieved the complete preservation of the existing network of arroyos and minimised earthwork alterations of the natural habitat. The building walls are made of rammed earth, recalling a tradition of indigenous desert building while meeting all of the performance requirements of modern use. The roof is covered in native desert cobble so that it blends into the desert when observed from the mountain trails to the east. The Gateway incorporates numerous strategies for resource conservation. An 18 KW solar system generates as much solar electricity as the Gateway consumes to realise a 'net zero' of energy consumption. Up to 60,000 gallons of rainwater is harvested through roof collection and storage in an underground cistern – providing 100% of the water needed for landscape irrigation.

项目设计在最小化对原生沙漠影响的前提下，突出了总面积36,400英亩（约14,730）的麦克道尔·索诺兰保护区的入口。项目是保护区内45英里（约72.4公里）长的徒步、自行车和骑马走道的入口。项目场地设计完全保护了原有的小溪，并且将土木工程建设对自然栖息地的影响降到了最低。
建筑的墙壁由素土夯实建成，让人回想起一种传统的本土沙漠建筑，同时也满足了所有现代的功能需求。屋顶上覆盖着沙漠卵石，从东面的山道上看，建筑似乎已经与沙漠融为一体。项目融合了多种资源保护策略。18千瓦的太阳能系统将为建筑提供足够的电力，使其符合"零能源"消耗。屋顶能够收集60,000加仑的雨水，并将它们储存在地下水池中，能够保证100%的景观灌溉用水。

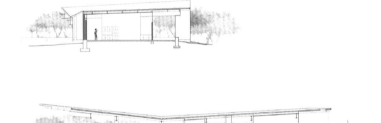

Awards:
2009 AIA Western Mountain Region Merit Award
2010 AIA Arizona APS Energy Award
2010 AIA Arizona Merit Award
USGBC LEED Platinum Certification

获奖情况：
2009 美国建筑师协会西山区域优秀奖
2010美国建筑师协会亚利桑那州分会APS能源奖
2010美国建筑师协会亚利桑那州分会优秀奖
美国绿色建筑委员会绿色建筑白金认证

1. The project minimised earthwork alterations of the natural habitat.
2. A night view of the architecture
3. The building walls are made of rammed earth
4. A night view of the façade
5. At night, the solar system could generate enough power for the architecture.
6. Entrance plaza

1. 项目土木工程建设对自然栖息地的影响降到了最低
2. 建筑夜景
3. 建筑的墙壁由素土夯实建成
4. 建筑立面夜景
5. 夜晚，太阳能系统将为建筑提供足够的电力
6. 建筑入口广场

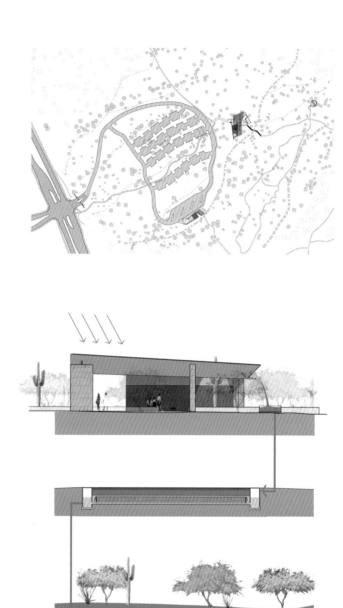

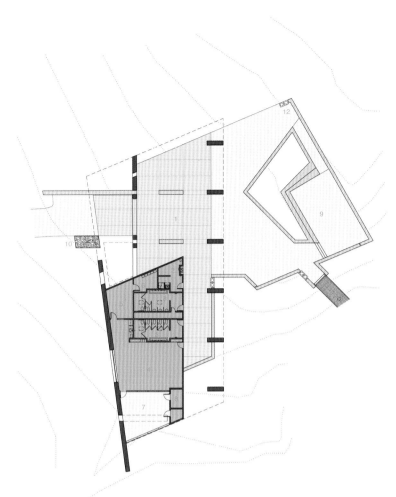

1. Desert cobble roof ballast
2. Threaded reinforcement bar
3. Concrete bond beam
4. Concrete seat beam
5. Densdeck roof substrate
6. Steel decking
7. R-30 batt insulation
8. 5/8 GYP.BD. ceiling
9. Rammed earth wall (95% excavated soil; 5% Portland cement)
10. Custom steel angle window
11. Conc.stem wall /window sill
12. Exposed ground concrete floor
13. Conc.footing

1. 沙漠卵石屋顶碎石
2. 螺纹钢筋
3. 混凝土结合梁
4. 混凝土座梁
5. 屋顶底板
6. 钢面板
7. R-30隔热层
8. 5/8 GYP.BD.天花板
9. 素土夯实墙（95%挖掘土；5%波特兰水泥）
10. 定制角钢窗
11. 混凝土墙/窗台板
12. 露石混凝土地面
13. 混凝土底脚

1. Entry plaza
2. Utility
3. Restroom
4. Open office
5. Office
6. Storage
7. Court
8. Mechanical
9. View terrace
10. Rain water catch basin
11. Pedestrian bridge
12. Recycle bins

1. 入口广场
2. 公共设施
3. 洗手间
4. 开放办公室
5. 办公室
6. 仓库
7. 庭院
8. 机械室
9. 观景平台
10. 雨水收集
11. 步行桥
12. 回收站

辰山植物园 Botanical Gardens Chenshan

Completion Date: 2010 **Location:** China, Shanghai **Gross Floor Area:** 58,600sqm **Designer:** Auer+Weber+Assoziierte **Photographer:** Auer+Weber+Assoziierte

竣工时间：2010年 项目地点：中国，上海 建筑面积：58,600平方米 设计师：奥尔+韦伯+合伙人建筑设计有限公司 摄影师：奥尔+韦伯+合伙人建筑设计有限公司

Architectural concept

The architectural elements are integrated into the continuum of the band-like 'ring of gardens'. The buildings strengthen the idea of the garden and formulate spots of concentrated experiences within the garden band; they are embedded into the undulating landscape and become a part of it. The dynamic forms and arrangement of the buildings and the change of materiality between concrete, as a supporting element of the landscape and glass that acts as a filling between the openings of the ring, integrate the architecture within the landscape design.

Positioning of the buildings

The emphasis of the architectural programme, for instance central entrance building, greenhouses and research centre, are integrated into the continuum of the 'garden ring' and positioned according to their particular function and relevance: central entrance building with visitor centre, education centre, exhibition area and administration on the south; research centre located between the research/experiment area and the botanical garden on the north constellation of greenhouses by the Shen Jing He canal, corresponding to the panoramic outline of the Chen Shan Mountains in the northeast.

建筑理念

建筑延续了园区内带状结构，强化公园设计理念的同时并将活动体验集中。建筑在造型、排列以及材质的变化（混凝土以及玻璃）上使其与景观设计融为一体，成为整个园区的一部分。

建筑结构排列

建筑主要功能结构，如中央入口大楼、研发中心以及花房，分别根据各自的功能及相互之间的联系排列：中央入口大楼包括游客中心、教育中心、展区以及行政区，布置在南侧；研发中心介于研发试验区及植物园之间，设计在北侧；花房设置在东北侧。

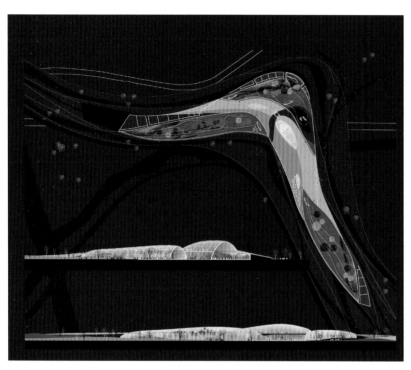

1. Visitor centre   1. 游客中心
2. Guest house   2. 宾馆
3. Research centre   3. 研究中心

1

4. Guest house
5. The architectural form, arrangement and the change of materiality integrate the architecture within the landscape design.
6. The visitor centre is positioned on the south.

4. 宾馆
5. 建筑的造型、排列以及材质的变化与景观设计融为一体
6. 游客中心布置在南侧

Floor plan level 2 二层平面图

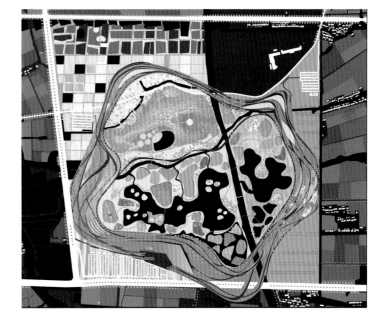

7. The visitor centre is made of concrete and glass.
8. Small plaza in front of the visitor centre
9. The research centre located between the research/ experiment area on the north
10. Greenhouses are located on the northeast.

7. 游客中心的材质是混凝土和玻璃
8. 游客中心前的小广场
9. 研发中心介于研发试验区及植物园之间，在北侧
10. 花房设置在东北侧

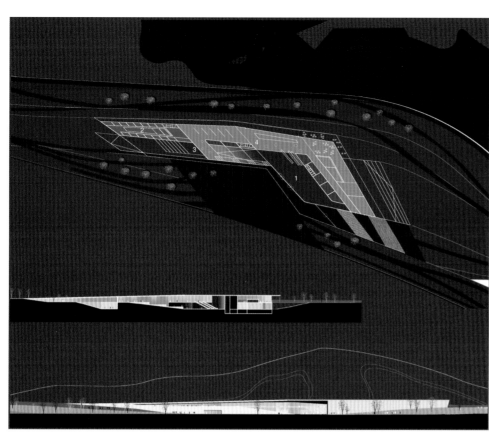

Botanical Gardens
1. Foyer
2. Administration
3. Court
4. Exhibition gallery

平面图
1. 门厅
2. 行政区
3. 庭院
4. 展览厅

11. Corridor to the entrance of the visitor centre
12. Entrance to the visitor centre
13. The high-ceiling lobby of the visitor centre
14. Wood stairs of the visitor centre

11. 通向游客中心入口的走廊
12. 游客中心入口
13. 举架高挑的游客中心大堂
14. 木质的游客中心楼梯

UNESCO World Natural Heritage, Messel Pit Visitor Information Centre

世界自然遗产——麦塞尔化石坑游客信息中心

Completion date: 2010 **Location:** Messel, Germany **Construction area:** 2,060 sqm **Designer:** landau + kindelbacher architekten **Photographer:** landau + kindelbacher/Jan Bitter

竣工时间：2010年 项目地点：德国，麦塞尔 建筑面积：2,060平方米 设计师：兰道+辛德尔巴切建筑事务所 摄影师：兰道+辛德尔巴切建筑事务所/詹·比特

The task of designing a visitor information centre for the Messel fossil pit, listed as a UNESCO World Natural Heritage site, demanded intense consideration of the turbulent history of the place, of both the scientific origin and the changing history of the site itself.

The stratification of the oil shale as genius loci forms the basic graphical idea of the building design. Like an earthen clod, the building breaks loose of the existing angular retaining walls and, with its significant monolithic wall panels, is oriented towards the pit – the actual highlight of the place. This movement culminates in the overhanging observation platform, from which one has an overview of the research site. The visitor symbolically wanders through the Earth's strata and enters a unique spatial structure that offers no analogy to conventional building types.

The building itself is perfectly tailored to the exhibition with its specific requirements. The various exhibition rooms in their overall appearance prepare the visitor for the subject matter that they deal with. This is achieved by simple but effective architectural means such as confinement and expanse, light and dark effects, high and low ceilings. The choice of materials was consciously restricted to sober and restrained to avoid interfering with a free interplay between changing scenographic installations. At the same time, a building has been erected that, through its design and the created atmospheres, blots out everyday life and makes a lasting impression.

The landscape architecture and the scenography take up and develop the architectural concept. For example, materials found in the theme garden and the outer areas of the site, such as oil shale, breezeblock fragments, or residues of oil production, are integrated in the presentation, as well as plants that characterised the appearance of the pit in earlier times. The exhibition architecture intensifies the atmosphere through its choice of colours and materials, as well as diversely designed exhibits specially developed in close cooperation with fellow researchers and scientists.

世界自然遗产——麦塞尔化石坑游客信息中心的设计要求设计师对场地动荡的历史进行充分的考虑，其中包括它的科学起源和场地自身变化的历史。

作为当地特色——油页岩的分层为建筑设计提供了主要平面概念。正如岩层一样，建筑摆脱了原有的倾斜式挡土墙，以其意味深长的单块墙板朝向化石坑——场地真正的亮点。这种运动以观景平台为终点，人们可以在平台上欣赏研究场地的全景。游客象征性地在地球地层里行走，进入与传统建筑截然不同的独特空间结构之中。

建筑是为特殊展览量身定制的。各种各样的展览厅为游客们提供化石坑相关的主题知识。限制和扩张、光影效果、高地错落的吊顶等简单而有效的建筑方式共同实现了这种效果。建筑师选择了素净的材料，避免装饰与变换的布景装置相互影响。同时，建筑通过独特的设计和氛围将人们带离了日常生活，给人留下了深刻的印象。

景观建筑与布景设计对建筑概念起到了至关重要的作用。例如，主题花园和场地外围的材料——油页岩、煤渣碎块和采油残留物——以及化石坑早期的特色植物都被纳入了保护之中。展览建筑通过色彩和材料的选择以及各种各样的展览设计（建筑师特别与相关研究人员和科学家进行了紧密的合作）凸显了场地的考古氛围。

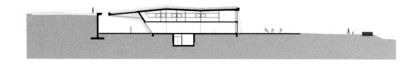
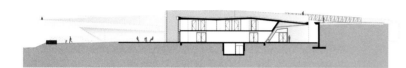
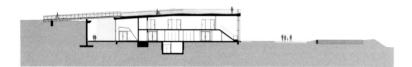
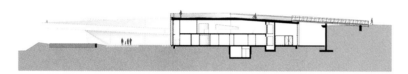
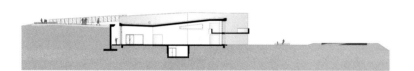

1. The geometric paving consists of oil shale, coal cinder, residue of oil extraction and feature plants.
2. Stairs to the roof
3. Visitors can have an overview of the site from the observation platform.
4. The building's façade

1. 几何状的地面由油页岩、煤渣碎块和采油残留物以及特色植物组成
2. 通向楼顶的楼梯
3. 观景平台上可以看到研究场地的全景
4. 建筑立面

Awards:
2011 Representative Buildings in state of Hessen, special recognition
2011 nominated for Mies Arch, European Union Prize

获奖情况：
2011德国黑森州代表建筑特别认证
2011密斯建筑欧盟奖提名

5. The geometric entrance and the paving form a unique architectural language.
6. The design concept reflects the stratification of the oil shale.
7. The pit with significant monolithic wall panels is the actual highlight of the place.

5. 几何状的建筑入口和入口的地面形成独特的建筑语言
6. 建筑设计理念体现了油页岩的分层的概念
7. 单块墙板朝向化石坑是场地的亮点

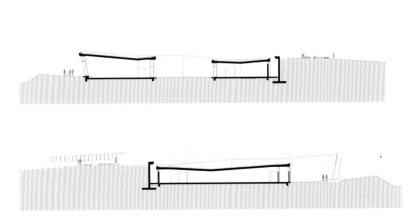
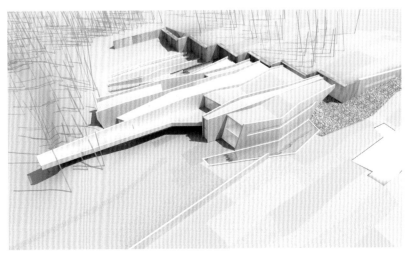

8. Roofs with different heights
9. Inside the retaining walls, the visitor symbolically wanders through the Earth's strata and enters a unique spatial structure.
10. Details of the entrance

8. 观景平台的建筑结构
9. 高低错落的屋顶
10. 进入挡土墙的游客好像在地球地层里行走，进入独特空间结构之中

1. Entrance	8. Earth's core room	1. 入口	8. 地心厅
2. Catering	9. Jungle	2. 餐饮区	9. 丛林
3. Sanitary facilities	10. Courtyard	3. 卫生设施	10. 庭院
4. Lobby	11. Preparation	4. 大厅	11. 准备室
5. Map room	12. Treasury	5. 地图室	12. 金库
6. Cinema	13. Platform	6. 放映室	13. 平台
7. Volcano room		7. 火山厅	

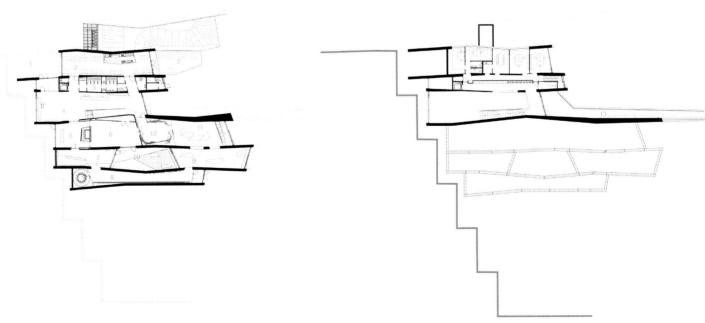

Tibet Linzhi Namchabawa Visitor Centre

西藏林芝南迦巴瓦接待站

Completion date: 2008 **Location:** Tibet, China **Site area:** 10,000 sqm **Construction area:** 1,500 sqm **Designer:** standardarchitecture **Structural System:** Mixed – Stone Masonry and Concrete Reinforcement, Concrete Roof

竣工时间：2008年 项目地点：中国，西藏 占地面积：10,000平方米 建筑面积：1,500平方米 设计师：标准营造 结构系统：混合——砌石和钢筋混凝土、混凝土屋顶

The Namchabawa Visitor Centre is the second building designed by standardarchitecture in Tibet. The building which was finished after the completion of the Yaluntzangpu Boat terminal is located in a small village called Pai Town in the Linzhi area, a town in the southeastern part of the Tibet autonomous region. The building sits on a slope along the road leading to the last Zhibai village deep in the Grand Canyon of Yaluntzangpu, facing the Yaluntzangpu River to its north and with the impressive 7,782-metre-high Mount Namchabawa at its background in the east.

The Namchabawa Visitor Centre serves as a welcome hub for visitors while providing them with comprehensive information about Mount Namchabawa and the Yaluntzangpu Grand Canyon. Simultaneously, the 1,500 m^2 building acts as the 'town centre' for the villagers as well as the supply base for the hikers exploring the canyon.

The programme of the building includes a reception/information hall, public toilets, a supply store, an internet bar, a medical centre, locker room for backpackers, meeting rooms, offices for tour guides and drivers, a water reservation tank and a central electrical switch house for the village.

Like a group of rock slices extending out of the mountain the Namchabarwa Visitor Centre consists of a series of stone walls set into the slope. The sharp geometry of the stone volumes is accentuated by an absence of windows in the wall facing the incoming road in the west, giving the building the appearance of a scale-less abstract element in the natural landscape.

Looked from afar the building neither hides itself, nor stands out from its background as a piece of 'Tibetan' Architecture. Approached from a distance on the road, people can't be sure if the volumes form a building or a set of retaining walls or even a 'Mani' wall at the foot of the mountain.

Taking off their cars, visitors follow a pathway led by a stone retaining wall up the hill, where they find the main entrance to the reception/exhibition hall. The main hall is lit by skylights and has a view window facing north towards the village and the Yaluntzangpu River. Entering the second layer of the one metre thick stone wall, visitors find the public toilets and the public luggage storage room; going further through another layer of stone wall they find the internet café, medical clinic and the driver's rest place. Halfway they have the choice of taking the 'stairway to heaven' up to the upper floor for the roof garden and the meeting rooms. The water tank is hidden beneath the stairs and the electric switch room takes the underground space.

The building is a mixed structure of traditional masonry stone walls and concrete. The walls are all made of local stone and vary in width of 60cm, 80cm and 1 metre. The constructional columns, ring beams and lintels inside the self-supporting stone walls reinforce the whole earthquake function of the walls, while supporting the concrete roof above them. The Tibetan craftsmen who built the stone walls are mainly from Shigatse. Their special customs and methods of building stone walls are not what can be designed on paper.

There are no popular Tibetan decorations either on the windows and doors of the building or in the interior space. It is obvious that Namchabawa Visitor Centre is contemporary architecture. The unique local characters embodied in this building should come from its local building materials and its unornamented construction process rather than disguises.

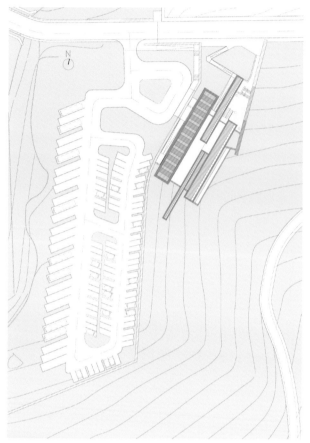

Site plan
总平面图

1. Looked from afar the building neither hides itself, nor stands out from its background.
2. Well-proportioned building
3. The centre consists of a series of stone walls set into the slope.

1. 从远处看，建筑既不会隐藏自己，又不会突出于背景
2. 错落有致的建筑
3. 建筑由一系列嵌入高坡的石墙组成

南迦巴瓦接待站是标准营造继西藏东南部林芝县的雅鲁藏布江小码头之后，在西藏建造的第二座建筑。建筑坐落在沿着通往雅鲁藏布大峡谷深处芝柏村的公路旁的高坡上，北临雅鲁藏布江，东倚海拔7,782米的南迦巴瓦峰。

作为一个欢迎游客的中心，南迦巴瓦接待站为他们提供南迦巴瓦峰和雅鲁藏布大峡谷的综合信息。同时，这座1,500平方米的建筑还是村民的"村中心"，也是探索大峡谷的背包客的供给基地。

建筑内设有前台/信息大厅、公共洗手间、供给仓库、网吧、医疗中心、背包客更衣室、会议室、导游和司机办公室、清水池和村庄中心变电房。

宛如一组沿着山峰伸展的岩片，接待站由一系列嵌入高坡的石墙组成。西面朝向公路的墙壁上没有开窗，突出了石结构犀利的造型，让建筑看起来像是自然景观中一件抽象的元素。

从远处看，建筑既不会隐藏自己，又不会突出于背景。从远处的公路上看，人们不能确定这是一座建筑、一组挡土墙还是山脚下的摩尼墙。

走下汽车，游客们沿着石墙旁的走道通过主入口进入前台/展览厅。大厅利用天窗采光，拥有一面朝向村落和雅鲁藏布江的观景窗。走进第二层一米厚的石墙，游客会发现公共洗手间和行李寄存室；再深入另一层石墙，是网吧、饮料站所和四季休息室。在半路上，他们有机会通过"天梯"进入二楼的屋顶花园和会议室。水池隐藏在楼梯下，而变电室则设在地下。

建筑是传统砌石结构和混凝土的结合物。墙壁全部由本地石材建成，厚度从60厘米、80厘米到1米不一。自我支承的石墙内的结构立柱、环形横梁和门窗过梁强化了墙壁的抗震功能，也支撑了上方的屋顶。建造石墙的西藏工匠主要来自日喀则地区。他们独特的习俗和建造方式是图纸所不能表示的。

无论是建筑的门窗还是室内空间都没有流行的西藏装饰元素。很明显，南迦巴瓦接待站是一座现代建筑。建筑所蕴含的地方特色来自于本地建筑材料和它朴素的建造流程，而不是简单的伪装。

4. The walls are all made of local stone and vary in width of 60cm, 80cm and 1 m.
5. The centre looks like a group of rock slices extending out of the mountain.
6. Simple architecture and zigzag paths

4. 墙壁由本地石材建成，厚度为60厘米、80厘米和1米
5. 建筑宛如一组沿着山峰伸展的岩片
6. 质朴的建筑与折线形的小路

Roof plan
屋顶平面图

Ground floor plan
首层平面图

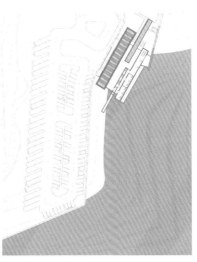

First floor plan
二层平面图

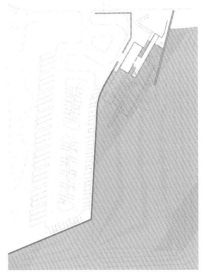

Basement floor plan
地下室平面图

7. The geometry of the stone volumes is accentuated by an absence of windows.
8. Local stone at the entrance and that on the wall have different forms.
9. The reception is lit by the skylights.
10. The entrance is an integrated mass.
11. The heavy stone walls divide various areas.

7. 墙壁上没有开窗,突出了石结构的造型
8. 入口处的本地石材在地面和墙壁的不同形态
9. 接待台利用天窗采光
10. 整个入口浑然一体
11. 厚实的石墙分割不同的区域

Queens Theatre in the Park

女王剧场

Completion date: 2011 **Location:** New York, USA **Site area:** 1,077 sqm **Designer:** Caples Jefferson Architects **Photographer:** Nic Lehoux, Julian Oliva

竣工时间：2011年 项目地点：美国，纽约 占地面积：1,077平方米 设计师：凯波斯·杰弗逊建筑事务所 摄影师：尼克·勒乌、朱利安·奥利弗

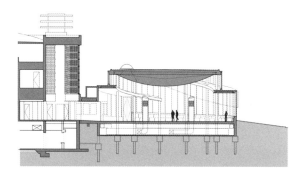

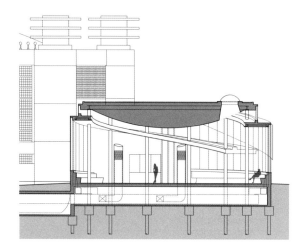

This is a place of reconnections, additions that build upon the playful circular geometries of the original 1964 Philip Johnson World's Fair complex. The new nebula room is a transparent viewing pavilion from which to appreciate the park's dreams of futures past, the Unisphere and the Johnson observatory towers and pavilions still in search of rescue. It is a party room for the Borough, whose rich materials and sunset colours are understood as festive by a wide cross-section of the 109 ethnic cultures that are the glory of Queens.

The new structure is a 600-person reception room for the Borough, standing on axis with the giant oval of Johnson's New York State Pavilion. The challenge was to create the impression of round spiraling forms while living within a budget that permitted only large flat glazed units. Using the principles of Gestalt psychology and the art of perspective, the architects designed a structurally glazed wall with metal fins projecting at each vertical joint. The effect of the fins is to drive the mind's eye to focus on the vanishing perspective that results from seeing them vanish around the curve.

The spiraling slope of the 'horizontal' mullions further intensifies this perception of curved movement in space. The design of the curtain wall utilises a broad array of contemporary technologies including low emulsion coatings to reduce solar heat gain, silicone sealant joints in lieu of metal mullion caps, gas-filled insulating units to reduce heating costs, and laminated glass outer lights to increase unit size and provide vandalism resistance. Digital design and fabrication techniques enabled fabrication of over 5,000 separate and unique glass panels that work in concert to create the illusion of perfect roundness.

Municipal Arts Society Masterworks Award 2011 citation: The renovated Queens Theatre in the Park is a transparent curved wall pavilion and a major addition to the iconic 1964 World's Fair complex designed by Philip Johnson. The compelling new structure provides a 600-person reception room and, significantly, fits within the constraints of the publicly funded city bid/build process. The project gives a shot of adrenaline to the park and energises this central part of Queens.

项目建在由菲利普·约翰逊设计1964世界展览会园区内，采用了有趣的圆形造型。全新的星云状空间是一个透明的观景亭，人们可以在此欣赏园区过去的未来之梦、巨型地球仪和约翰逊观景塔。作为这一区域的联谊室，其丰富的材料和落日系色彩在女王统治的109种民族文化中都是具有喜庆意义的。

新结构是一个可容纳600人的接待厅，位于约翰逊纽约州馆巨大椭圆区域的中轴上。项目面临的挑战是利用仅够制作大片平面玻璃的预算来打造圆形螺旋造型。建筑师利用格式心理学理论和美学观点，以向外辐射的垂直金属翅片来制造玻璃幕墙。翅片的效果让人将注意力放在消失的远景上，让它们看起来像是沿着圆弧消失了一样。

水平窗框的螺旋斜率进一步凸显了空间的弧形运动。幕墙设计采用了大量现代技术，其中包括：以低挥发乳胶涂层来减少热增量；以硅酮密封剂替代金属窗框帽；以空气填充隔热单元来减少供暖成本；以夹层玻璃来增加单元尺寸、增强耐久性。数字设计和装配技术让5,000多个独立玻璃板得到了统一装配，以达到完美的圆形错觉。

引自"2011市政艺术社团大师奖"颁奖词：整修后的女王剧场是在一座透明的圆形亭，建于由菲利普·约翰逊设计1964世界展览会园区内。引人注目的新结构提供了可容纳600人的接待厅，十分符合公共集资的城市投标/建造流程。项目是园区的催化剂，让它重新获得了活力。

1. Main entrance
2. Gas-filled insulating windows
3. A bird's eye view of the interesting rounded architecture
4. The transparent pavilion
5. The large flat glazed units create the round spiraling forms

1. 建筑入口
2. 夹层玻璃窗
3. 俯视有趣的圆形建筑
4. 透明的观景亭
5. 大片平面玻璃打造圆形螺旋型的建筑

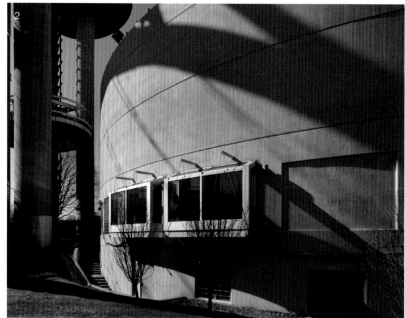

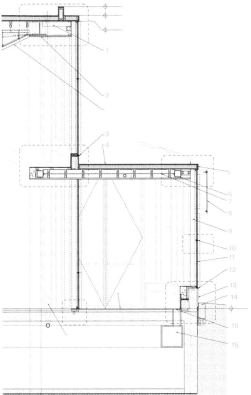

1. Ts 44, welded to beam @ underside of top flange
2. Pigmented PL skimcoat over 5/8" GWB
3. Curved aluminium sill, slope: 4 3/4" rise per mullion bay
4. TZAC copper flat seam roofing & fascia
5. TZAC copper fascia
6. TS 66
7. W413
8. Metal screen supported by steel mullion
9. 3" 4 PTD. steel mullion tube to support aluminium glazing system, prefinished
10. 1/4"2" alumin. bar btwin glass break
11. Curtain wall: double glazing
    W/PVB laminated glass attached to alumn. system by structural silicon
12. Curved aluminium sill W/drip edge slope: 1" rise per mullion bay
13. 1" acrylic stucco over 7 1/2" cast-in-place concrete
14. Conc. mowing strip
15. Supply air duct

1. Ts 4×4，焊接在梁上，上翼缘的底面
2. 彩色石膏灰泥涂层，在5/8"石膏墙板之上
3. 弯曲铝窗台，坡度：每个窗框上升4 3/4"
4. TZAC铜平缝屋顶和横带
5. TZAC铜横带
6. TS 6×6
7. W4×13
8. 由钢框支撑的金属网
9. 3"×4 PTD.钢框管，支撑铝制玻璃装配系统，预加工
10. 玻璃缝隙之间的1/4"×2"铝条
11. 幕墙：双层玻璃
    W/PVB夹层玻璃通过结构硅胶附着于铝系统之上
12. 弯曲铝窗台，滴水槽檐坡度：每个窗框上升4"
13. 1"丙烯酸灰泥，覆于7 1/2"现场浇筑混凝土之上
14. 混凝土带
15. 供气管道

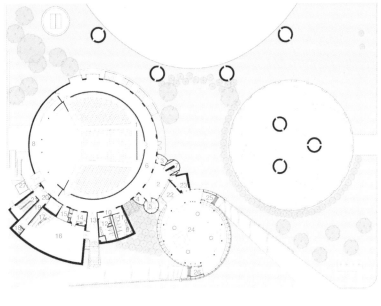

1. Theatre	1. 剧院
2. High lobby/safe area	2. 大厅/安全区
3. Concession	3. 特许区
4. Manager's office	4. 经理办公室
5. Elevator 1	5. 电梯1
6. Lobby corridor / safe area	6. 大堂走廊/安全区
7. Stair 1a	7. 楼梯1a
8. Backstage	8. 后台
9. Janitor closet	9. 门卫室
10. Toilet vestibule	10. 洗手间前厅
11. Female public toilet	11. 女公共洗手间
12. Male public toilet	12. 男公共洗手间
13. Cabaret lobby / safe area	13. 卡巴莱大堂/安全区
14. Food service	14. 食品服务
15. Lighting control room	15. 灯光控制室
16. Cabaret	16. 歌舞表演厅
17. Bar	17. 酒吧
18. Actor toilet	18. 演员洗手间
19. Corridor	19. 走廊
20. Stage vestibule	20. 舞台前厅
21. Stair 2	21. 楼梯2
22. Low lobby	22. 低层大堂
23. Box office	23. 售票处
24. Entry lobby / safe area	24. 入口大厅/安全区
25. Vestibule 1	25. 门廊1
26. Vestibule 2	26. 门廊2
27. Loading dock	27. 装货码头

6. Interior entrance
7-9. The rich materials and sunset colours highlight the festive meaning.

6. 室内入口
7-9. 室内设计采用丰富的材料和落日系色彩突出喜庆意义

Visitors Pavilion at Ramat Hanadiv

莱姆特·哈那迪夫游客亭

Completion date: 2008 **Location:** Zichron Ya'akov, Israel **Area:** 2,700 sqm **Designer:** Ada Karmi-Melamede Architects **Photographer:** Amit Geron **Client:** Yad Hanadiv Foundation

完成时间：2008年 项目地点：以色列，吉士隆雅科夫 项目面积：2,700平方米 设计师：阿达·卡米–麦拉梅德建筑事务所 摄影师：阿米特·杰隆 委托人：亚德·哈那迪夫基金会

This building is located between a vast parking area which is designed to serve the public and Ramat Hanadiv Memorial Gardens. On 12th March, 2008, Ramat Hanadiv's Visitors Pavilion became the first building in Israel to be granted standard certification for sustainable construction. The Visitors Pavilion was planned by architect Ada Carmi-Melamed around two perpendicular axes. The landscape axis connects the Memorial Gardens and the Nature Park, while the horizontal axis, in the form of a rounded, green hill, envelopes the entrance plaza to the Gardens.

The architectural concept evolved from the need to create a foyer in front of the gardens which will function as a meeting place for the diverse groups of visitors.

The architectural composition extends approximately 150 metres in a curvilinear fashion. It contains an auditorium, classrooms, small courtyards, an exhibition space and a small cafeteria at the extremity. These functions are housed within a berm formed by two inclined landscaped surfaces that lean against each other with light penetrating in between.

A pedestrian path, lit by indirect reflected light from above, stretches the entire length of the berm connecting all the various functions.

Three passages cross the berm and connect the parking area with the open foyer. The central passage faces directly onto the gate of the Memorial Gardens and joins its formal axis. A landscape of tranquility and splendour embraces the axis.

建筑坐落在一片广阔的停车场上，停车场原本是为公众和莱姆特·哈那迪夫纪念园所设计的。2008年3月12日，莱姆特·哈那迪夫游客亭成为了以色列第一座获得可持续建造标准认证的建筑。游客亭由建筑师阿达·卡米–麦拉梅德围绕两条垂直的轴线设计。景观轴线与纪念园和自然公园相连，而水平轴线则以圆形的山丘为造型，将纪念园的入口广场包围起来。

建筑概念源于为纪念园打造一个前厅的需求，为各种游客群体提供一个集会场所。

建筑以曲线方式扩展约150米。它包含礼堂、教室、小型庭院、展览空间和小型自助餐厅。这些功能区全部设置在由两个倾斜景观坡面组成的坡台之中，两个坡面相互倚靠，其间设有光缝。

人行道利用上方的间接反射光进行照明，延伸了整个坡台，将各个功能区连接起来。

三条走廊贯穿坡台，连接了停车场和开放门厅。中央的走道正对纪念园的大门，与中轴合并在一起。宁静而华美的景观环绕着设计轴线。

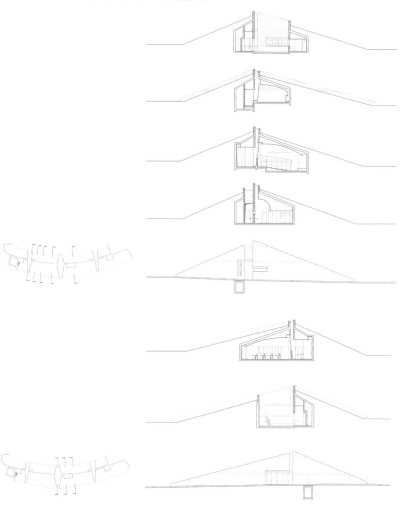

1. Entrance to the gardens
2. Building silhouette from the parking
3. Offices Pavilion

1. 花园入口
2. 从停车场看建筑的轮廓
3. 办公楼

Awards:
2009 AI Award
2010 Best Design by DOMUS Israel

获奖情况：2009年AI奖
2010年以色列DOMUS最佳设计

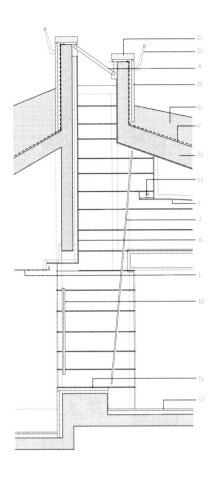
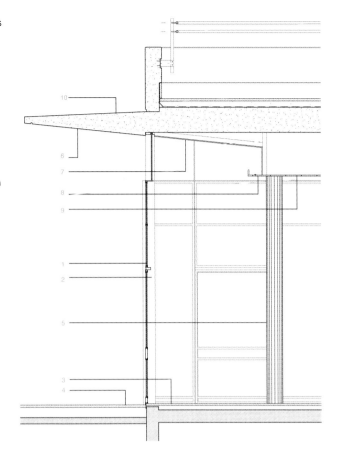

A. Steel skylight with Extra-Clear Low-E glass panels
B. 40mm stone facing
C. 100mm coping stone
D. Painted 40 mm metal railing
E. Soil (Gardening Mix)
F. 20mm Insulation and Vapor burrier layers
G. 350mm tilted concrete ceiling
H. Indirect T5 lighting
I. Plaster board ceiling
J. 14mm reinforced glass panel
K. Plaster board double wall
L. Hung wooden acoustic ceiling
M. 2 18mm glass layers with decorative film betwin
N. 50mm stone coping
O. 20mm Interior polished stone floor

A. 钢框天窗，配有超清低辐射玻璃板
B. 40mm石料镶面
C. 100mm压顶石
D. 40mm彩色金属栏杆
E. 土壤（园艺混合）
F. 20mm隔热隔气层
G. 350mm倾斜天花板
H. 间接T5照明
I. 石膏板天花板
J. 14mm加固玻璃板
K. 石膏板双层墙面
L. 悬挂式木制隔音天花板
M. 双层18mm玻璃，中间夹有装饰膜
N. 50mm压顶石
O. 20mm室内抛光石地面

4. Berm's western edge　4. 坡台西端
5. Open foyer from east　5. 从东侧看开放式门厅
6. Secondary passage　6. 二级走道
7. Open foyer from east　7. 从东侧看开放式门厅

1. Steel mullions and Extra-Clear Low-E glass panels　1. 钢框和超清低辐射玻璃板
2. Vertical load-burying steel mullions　2. 垂直荷载钢框
3. 20mm Interior polished stone floor　3. 20mm室内抛光石地面
4. 30mm Exterior Bush-hammered stone paving　4. 30mm室外荔枝面石铺面
5. Constructive 250 mm painted steel columns　5. 250mm结构涂漆钢柱
6. Cast in-situ concrete cantilever　6. 现场浇筑混凝土悬臂
7. Plaster board ceiling　7. 石膏板天花板
8. Indirect LED lighting　8. 间接LED照明
9. Hung wooden acoustic ceiling　9. 悬挂式隔音木天花板
10. External transparent insulation　10. 外部透明隔热层

8. Main passage
9. Light and shade in secondary passage
10. Kiosk and internal courtyard
11. Ramp to internal courtyard

8. 主走道
9. 二级走道的光影效果
10. 凉亭和内庭
11. 通往内庭的坡道

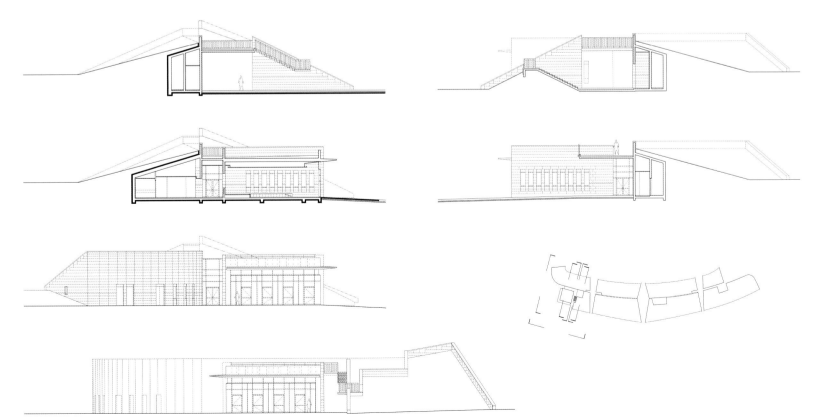

12. Classroom
13. Auditorium
14. Internal walkway ramp
15. Natural daylight on WC

12. 教室
13. 礼堂
14. 内部步行坡道
15. 洗手间采用自然光

14

15

1. Entry
2. Auditorium
3. Classrooms
4. Courtyard
5. Exhibition hall
6. Lecture room
7. Courtyard
8. Cafeteria
9. Internal walkway
10. North-South pedestrian passages

1. 入口
2. 礼堂
3. 教室
4. 庭院
5. 展览厅
6. 演讲厅
7. 庭院
8. 自助餐厅
9. 内部走道
10. 南北向步行走廊

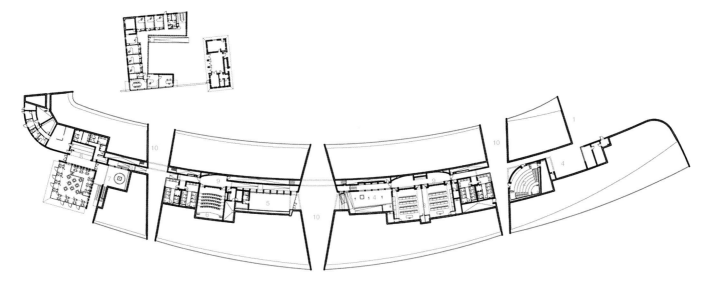

Urban Redevelopment of the Plaza del Milenio

千禧广场城市改造

Completion date: 2011 **Location:** Valladolid, Spain **Site Area:** 25,000 sqm **Designer:** EXP architects **Photographer:** TAFYR, EXP architects

竣工时间：2011年 项目地点：西班牙，巴利亚多利德 占地面积：25,000平方米 设计师：EXP建筑事务所 摄影师：TAFYR、EXP建筑事务所

The project of urban redevelopment by EXP architects consists in renewing the square, with its new pavilion in the immediate environment, and providing for a multiuse structure with a capacity of 1,500 persons.

Valladolid, an historic city of 320,000 inhabitants, located in North West Spain, is the capital of the province of Castile and Leon. The city stands at the confluence of the Pisuerga and the Esgueva rivers. The district of the Plaza del Milenio borders the right bank of the Pisuerga and is connected to the historic centre by the Isabel the Catholic Bridge.

The project concerns the development of public spaces. It consists in re-connecting the district with the city centre and of enhancing the banks of the Pisuerga river.

The project comprises several interlinked components: a car park, a pavilion dedicated to cultural events, a vast square, and an environmental reserve. The project addresses the development of the area at every scale, from the street furniture to the landscape of the river itself.

To extend the district beyond the bridge towards the city, the project is based on a progressive expansion of the ecosystem of the river towards the public space. It places the pavilion at the very heart of a new green space connected to the city center thanks to the renovation of the Isabel the Catholic Bridge.

The ground of the square lifts up slightly, providing an artificial topography of turfed slopes in continuity with the movement of the banks, allowing the atmosphere of the river to migrate towards the urban space. By a gradual passage from the vegetable to the mineral, nature is thus tamed, domesticated on the square and all around the pavilion through the creation of thematic gardens, water gardens, and the planting of remarkable trees.

The banks of the river are treated as an urban park offering new assets: observation platforms at the water's edge, collective and individual street furniture, interactive children's games, mist generators and water fountains, scenic geysers, sculptures, and lighting highlighting contour lines.

The role of the Isabel the Catholic Bridge is reinforced by a complete upgrade to provide for modern urban needs. It is equipped with bridges for pedestrians and cyclists, decked in wood, and enhanced by a lighting device system coordinated with that of the square, partly fed by vertical wind turbines and photovoltaic panels fixed to the parapet. Thus retrofitted, the bridge becomes a full element of the new urban development, constituting part of the energy infrastructure, consistent with the environmental aspirations of the project.

A series of ecological and sustainable processes is featured on the site as a whole: clean construction processes, manufacturing process, recyclable materials, amorphous solar panels, LED lighting system, vertical urban wind turbines, rainwater recycling, planting of rustic native species, recharging terminals for electric vehicles, self-service bikes, etc.

The project is conceived as a multifunctional system, a vibrant, inspiring place, drawing people together, strengthening local identity and enhancing the quality of the environment, all the while providing a capacity for evolutionary change.

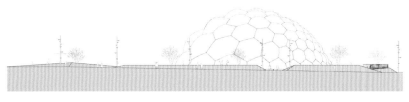

1.2. Rest area in the plaza
3. A bird's eye view of the architecture and landscape
4. View the architecture from Pisuerga River

1、2. 广场上的休息区
3. 建筑与景观俯视图
4. 从皮苏埃加河看多功能建筑

EXP建筑事务所进行的城市改造项目包括以新建的凉亭复兴广场和提供一个可容纳1,500人的多功能结构。

作为一个拥有320,000居民的古城，巴利亚多利德坐落在西班牙西北部，是卡斯提尔和利昂的省会。城市位于皮苏埃加河和艾斯吉瓦河的交汇处。千禧广场区毗邻皮苏埃加河的右岸，通过伊莎贝拉一世大桥与城市的历史中心相连。

项目注重公共区域的开发，包括重新连接这一区域与市中心和提升皮苏埃加河两岸的环境条件。

项目由若干相互连接的元素组成：停车场、文化活动馆、宽敞的广场和环境保护区。项目在各个层面上对区域进行开发，从街道设施到河流景观。

为了让桥后方的区域延伸至城市，项目以河流生态系统向公共空间的逐步扩张为基础。由于伊莎贝拉一世大桥得到了翻新，项目在每个连接市中心的绿地中心都设置了场馆。

广场的地面缓缓上升，形成了与河岸的运动相一致的草坡，让河流的气氛渗透到城市空间中。从植被到矿物，自然在广场上得到了驯服。主题花园、水上花园和纪念树木将场馆环绕了起来。

河岸被处理成提供新景点的城市公园：河畔观景台、组合和独立的街头装置、互动儿童游戏区、喷雾装置和喷泉、间歇喷泉、雕塑以及凸显轮廓线的灯光。

伊莎贝拉一世大桥的地位通过升级得到了提升，满足了城市需求。它配有行人和自行车专用桥，以木板铺面，并且装配了与广场相匹配的灯光系统。部分灯光电力由垂直风涡轮和栏杆上的太阳能电池板提供。经过翻新，大桥成为了城市开发中一件完整的元素，配有能源设施，并且与项目的环境愿望相一致。

项目拥有一系列完整的生态和可持续流程：清洁建设流程、制造过程、可循环材料、非晶质太阳能电池板、LED照明系统、垂直风涡轮、雨水收集、本土植物种植、电动车充电站、自助自行车等。

项目是一个多功能系统，一个活跃、振奋人心的空间，它将人们汇聚在一起，增强了区域形象并且提升了环境质量。最重要的是，它有潜力进行进一步的改造。

5. The unique lights are powered by vertical wind turbines and photovoltaic panels fixed to the parapet.
6. Lawn in the plaza
7. Scenic geysers
8. Isabel the Catholic Bridge is equipped with bridges for pedestrians and cyclists, decked in wood, and enhanced by a lighting device system coordinated with that of the square.

5. 广场上别具特色的灯,电力由垂直风涡轮和栏杆上的太阳能电池板提供
6. 广场上的草坪
7. 间歇喷泉
8. 伊莎贝拉一世大桥,配有行人和自行车专用桥,以木板铺面,并且装配了与广场相匹配的灯光系统

9.10. Main entrance
11.12. Night view of the architecture

9、10.多功能建筑入口
11、12.建筑夜景

11

1. Pavilion
2. Turfed slopes
3. Modular benches
4. Water garden
5. Aromatic plants
6. Amarelle tree
7. Pedestrian pathway
8. Access
9. Terrace
10. Fountain
11. Children gardens
12. Interactive fountain
13. Observation platforms

1. 凉亭
2. 草坡
3. 模块化长椅
4. 水景花园
5. 芳香植物
6. 酸樱桃树
7. 人行道
8. 入口
9. 平台
10. 喷泉
11. 儿童花园
12. 互动喷泉
13. 观景台

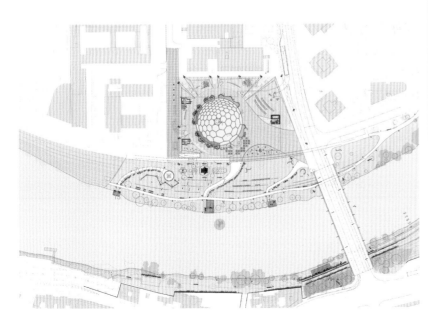

12

Red Rock Canyon Visitor Centre

红岩谷游客中心

Completion date: 2010 **Location:** Las Vegas, USA **Area:** 4,366 sqm **Designer:** Line and Space, LLC
Photographer: Robert Reck

竣工时间：2010年 项目地点：美国，拉斯维加斯 项目地点：4,366平方米 设计师：线与空间公司 摄影师：罗伯特·莱克

The community of Las Vegas is one of the most rapidly growing areas in the United States. The Mojave Desert is a delicate place. Extreme heat and little rain exaggerate the time it takes the land to recover from human disturbance. Educating this growing population on respectful living in the desert is a priority. The Bureau of Land Management (BLM), with support from the Red Rock Canyon Interpretive Association, the Friends of Red Rock Canyon, and the Master Gardeners has used the current Visitor Centre to apply their mission. After 28 years, even with numerous upgrades and additions, the facility no longer provides for the needs of both staff and visitors. Space is inadequate for users, exhibits, and storage. The building is increasingly expressing its age visually, and in wear and maintenance.

In support of the BLM's mission to encourage stewardship for the land, the design of the new facility provides an outdoor experience which will instill in individuals, a sense of personal responsibility for their land's well-being. Key to the visitor's understanding of Red Rock Canyon is the need to experience, and be a part of the inspirational desert landscape. The design of the new Visitor Centre fulfills this, where up to 1,000,000 visitors per year are introduced to the geology, science, art and culture of Red Rock Canyon, and encouraged to visit the nearby real thing.

The facility differs from traditional visitor centres by emphasising the specific attributes of Red Rock Canyon itself, in lieu of traditional dioramas and pseudo-natural imitations. Here, visitors are introduced to outdoor abstract sculptural exhibits designed to inform them about the surrounding landscape and prepare them for their own explorations within the National Conservation Area. This innovative approach moves visitors and interpretation from interior conditioned space to exterior microclimates which are kept comfortable through the use of shade, evaporative cooling, and a large energy efficient fan. During summer months, these strategies make an individual viewing outdoor exhibits feel approximately 10-15° cooler than being in direct sunlight. This shifting of exhibits from air conditioned interior space to fully day-lit tempered outdoor microclimates leads to massive energy savings for the client and a truly unique experience for visitors.

Many resource-conserving ideas are incorporated into the facility. The Arrival Experience is sheltered by a "big hat" (a roof with ample overhangs) which creates intermediate thermal transition zones and forms the collection plane for rainwater harvesting (used for interpretive exhibits and landscape irrigation). These thermal transition zones alleviate shock to users' senses by providing a place where eyes and skin can adjust while moving between the hot, bright, exterior and the cool, shaded interior. The overhangs also shade floor to ceiling glass in the summer while allowing the low winter sun to enter the space, reducing demand on the building's mechanical system.

High-efficiency mechanical systems are utilised: solar water heating, a transpired solar collector system and a 55 KW photovoltaic array convert the region's intense sun into free energy. The transpired solar collector provides heating for the Rest rooms, allowing the mechanical system in these spaces to be eliminated. Natural and durable materials such as concrete masonry,

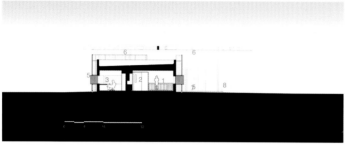

1. Break room
2. Staff entry
3. Staff office
4. Steel window box
5. Perforated steel screen
6. Overhead shade canopy
7. Existing toll booth
8. Toll lane

1. 休息室
2. 员工入口
3. 员工办公室
4. 钢窗盒
5. 穿孔钢屏
6. 遮阳凉亭
7. 原有售票亭
8. 售票车道

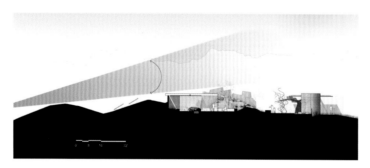

1. Transpired solar wall
2. Photovoltaics
3. Natural habitats
4. Entry

1. 太阳能墙
2. 太阳能光电板
3. 自然栖息地
4. 入口

steel and glass reduce maintenance needs and help unify the building with the landscape. As part of future upgrades to infrastructure, a new recirculating waste water system will replace an existing septic system, treating reclaimed water for reuse in flushing toilets.

The Visitor Centre is an asset for the community, offering creative interpretive presentations and a unique opportunity to interact with one of the most exceptional ecosystems in the United States. The building stands as a physical example of how to exist and conserve in the desert by extending the usability of outdoor space, providing ample shade, harvesting rainwater, generating its own energy, and using natural and durable materials. The hope is that visitors will realise the benefits that these systems can offer, and then apply some of the same concepts in their own lives.

拉斯维加斯是美国增长速度最快的地区之一。莫哈韦沙漠是一个脆弱的地方。异常的高温和极少的降水让土地从人为干扰中恢复的时间变得更长。教育人们如何尊重沙漠生活迫在眉睫。在红岩谷说明协会、红岩谷友好协会和园艺大师机构的支持下，土地管理局利用现有的游客中心来完成任务。28年后，尽管历经了数次升级和扩建，游客中心已经无法满足员工和游客的需求，没有足够的空间用以使用、展览和仓储。建筑在视觉上也逐渐衰败，也亟待维护保养。

为了支持土地管理局对土地的管理工作，新设施的设计提供了向人们灌输土地责任感的户外体验。游客理解红岩谷的关键在于体验——成为沙漠景观的一部分。新游客中心的设计实现了这点，每年，超过1,000,000名游客将会获得红岩谷地质、科学、艺术和文化相关的知识，并且对实物进行近距离参观。

项目与传统的游客中心不同，它突出了红岩谷的特质，而不是传统实景模型和虚假的自然模仿。在这里，游客可以通过露天抽象雕塑展来了解周边景观，获取探索自然保护区所需的信息。这种创造性方式把游客和说明从室内空间移到了户外微环境，并且通过遮阳设施、蒸发制冷和大型节能扇来保证舒适度。在夏季，这些策略让户外展览区比阳光直射区域的温度凉爽10~15摄氏度。将展览从室内空调环境转移到日光调和的露天微气候能够节省大量的能源，也为游客提供了独特的体验。

项目融入了许多节约资源的想法。入口处的"巨大帽子"（一个带有宽敞悬臂结构的屋顶）打造出过渡热转换区，也形成了雨水收集平面（雨水被用于说明展览和景观灌溉）。这些热转换区缓和了使用者的感官，让眼睛和皮肤可以在炎热明亮的室外和凉爽偏暗的室内之间得到调节。夏天，悬臂结构为落地玻璃窗提供了阴凉；冬天，它让低斜的阳光进入室内，减少了建筑机械系统的需求。

项目采用了诸多高效机械系统：太阳能热水器、太阳能手机系统和55千瓦的太阳能电池板阵列将当地强烈的阳光转化为免费的能源。太阳能收集器为洗手间提供了供暖，让这些空间无需多余的机械系统。混凝土砌块、钢铁和玻璃等自然而耐久的材料减少了养护需求，帮助建筑与景观融为一体。作为未来的升级设施，一个全新的废水循环利用系统将取代原有的系统，将循环水用于抽水马桶冲水。

游客中心是社区的宝藏，提供了创意十足的展示说明和与美国最独特的生态系统互动的好机会。建筑扩展了户外空间的利用率、提供了足够的阴凉、对雨水进行收集、自主发电并且利用自然和耐久的材料，堪称沙漠建筑的典范。建筑师希望游客能够认识到这些系统带来的便利，并且在生活中获得一些同样的概念。

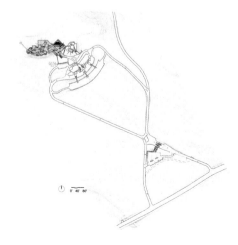

5. Contact Station
6. Fire Pavilion
7. Air Pavilion

5. 联系站
6. 火亭
7. 空气亭

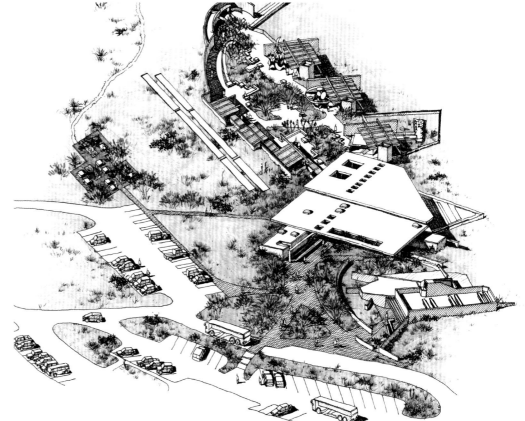

8. Gathering area
9,10. View deck

8. 集合区
9、10. 观景平台

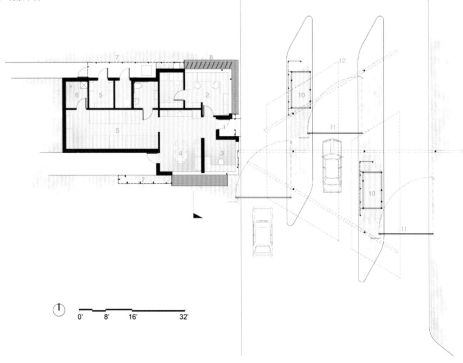

1. Staff entry
2. Main toll booth
3. Staff office
4. Break room
5. Storage
6. Janitor closet
7. Perforated steel screen
8. Vertical shade louvers
9. Steel windows box
10. Existing toll booth
11. Toll lane
12. Overhead shade canopy

1. 员工入口
2. 主售票亭
3. 员工办公室
4. 休息室
5. 仓库
6. 门卫室
7. 穿孔钢屏
8. 垂直遮阳百叶窗
9. 钢窗台盒
10. 原有售票亭
11. 售票车道
12. 遮阳凉亭

10

1. Entry plaza
2. Transition space / entry
3. Temporary exhibits / multipurpose
4. Information desk
5. Arrival experience
6. Panorama window
7. Classroom with outdoor patio
8. Gift shop
9. Transpired solar collector
10. Outdoor amphitheater
11. Earth pavilion
12. Tortoise habitat
13. Fire pavilion
14. Air pavilion
15. Four elements exhibit
16. 360 view deck
17. Cliff walls exhibit
18. Water walk
19. Natural habitats
20. Desert spring exhibit
21. Desert ecosystem exhibit
22. Water harvesting storage
23. Earth berms
24. 60kw photovoltaic array

1. 入口广场
2. 过渡空间/入口
3. 临时展览/多功能厅
4. 信息站
5. 到达体验
6. 全景窗
7. 配有露天天井的教室
8. 礼品店
9. 太阳能收集器
10. 露天剧场
11. 大地馆
12. 龟类栖息地
13. 火馆
14. 空气馆
15. 四元素展览
16. 360度观景平台
17. 崖墙展览
18. 水道
19. 自然栖息地
20. 沙漠泉水展览
21. 沙漠生态系统展览
22. 水收集库
23. 大地坡台
24. 60kw光电伏阵列

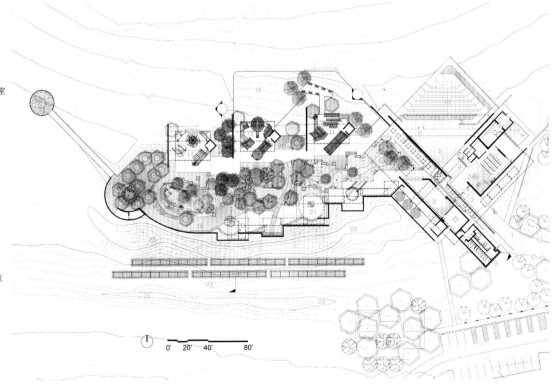

阿尔法城迎宾中心 Welcome Centre of Alphaville

Completion date: 2009 **Location:** Brasilia, Brazil **Site Area:** 11,000 sqm **Designer:** Luis Fernando Rocco, Fernado Vidal, Douglas Tolaine **Collaborators:** Daniela Cunha, Reinaldo Ferreira, Carolina Ishibashi **Photographer:** Daniel Ducci

竣工时间：2009年 项目地点：巴西，巴西利亚 占地面积：11,000平方米 设计师：路易斯·费尔南多·罗科、菲尔那多·维达尔、道格拉斯·托莱恩 合作设计：丹妮拉·库尼亚、雷纳尔多·费雷拉、卡罗里那·石桥

The bold, contemporary building stands out for the originality of the implementation of volume, which accompanies the shape and slope of the land of almost 11,000 square metres.
Designed to serve in a first moment as sales stand and Welcome Centre of Alphaville, which is a gated community away from Brasilia downtown, and later a permanent facility for visitors interested in the project and conviviality, also will host events and cultural activities. The blocks accompany the lot's format and declivity, and are united by a Corten steel grid composing the cover. The grid with two thousand square metres stands out in the complex due to its dimensions as well as for creating a central plaza which large trees were planted. This covering grid crosses a 30 metres span and sits on seven concrete blades supporting a load of 50 tons. At some points the grid presents large openings leaving room for the tree tops.

这座大胆而现代的建筑以其独具原创性的空间脱颖而出。它依坡地地势而建，总面积达11,000平方米。项目最初被设计为阿尔法城（远离巴西利亚市中心的一个封闭社区）的销售和迎宾中心，不久之后，它成为了一个永久性设施，为对项目有兴趣的游客提供了欢宴的场所，举办各色宴会活动和文化活动。各个楼体随着场地的地势而建，由一个柯尔顿钢格屋顶统一起来。栅格结构总面积2,000平方米，延伸出建筑之外，形成了一种植着树木的中央广场。屋顶横跨30米的跨度，下方的7面混凝土墙能够承重50吨。一些区域的栅格拥有巨大的开口，为树冠预留出足够的空间。

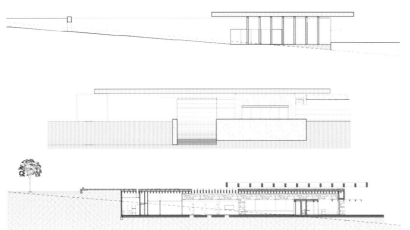

1. The architecture accompanies the shape and slope of the land.
2. The grid stands out in the complex.
3. The blocks are united by a steel grid composing the cover.
4. Seven concrete blades support the roof.

1. 建筑依坡地地势而建
2. 栅格结构屋顶延伸出建筑之外
3. 各个楼体由钢格屋顶统一起来
4. 屋顶下方有7面混凝土墙

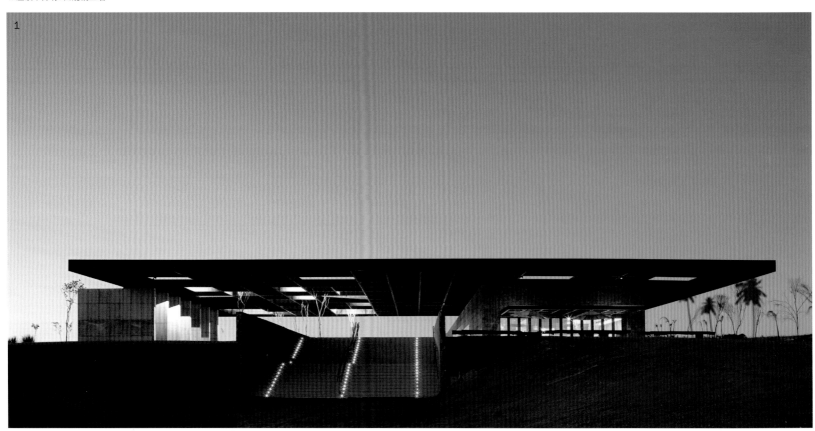

5. Central plaza with plants
6.7. At some points the grid presents large openings leaving room for the tree tops.

5. 种植着树木的中央广场
6、7.栅格有巨大的开口，为树冠预留空间

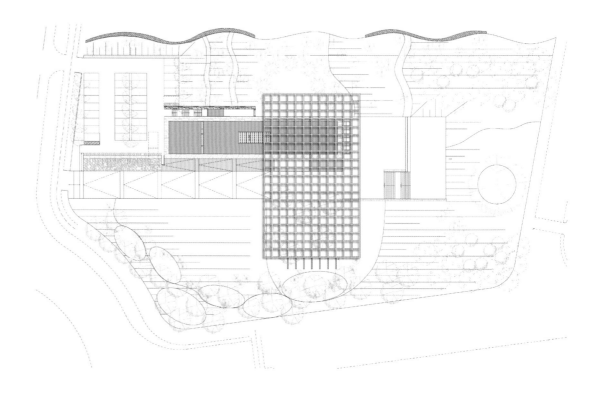

8. The plaza sets some facilities for conviviality.
9. 10. Interior spaces to host events and cultural activities

8. 广场上的游客欢宴场所
9、10. 举办宴会活动和文化活动的室内空间

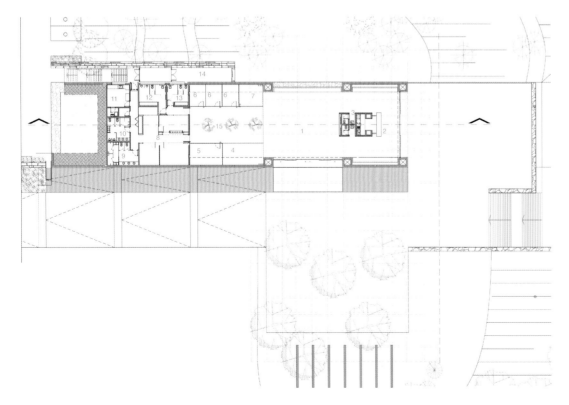

1. Main room
2. Coffee
3. Restrooms
4. Model
5. Kids room
6. Meeting room
7. Projection
8. Staff
9. Men's Dressing room
10. Women's Dressing room
11. Kitchen
12. Men's restroom
13. Women's restroom
14. Technical space
15. Courtyard

1. 主大厅
2. 咖啡厅
3. 洗手间
4. 模型
5. 儿童房
6. 会议室
7. 投影室
8. 员工休息室
9. 男更衣室
10. 女更衣室
11. 厨房
12. 男洗手间
13. 女洗手间
14. 技术空间
15. 庭院

9

10

Mirror House at Copenhagen Central Park

哥本哈根中央公园镜子屋

Completion date: 2011 **Location:** Copenhagen, Denmark **Area:** 140 sqm **Designer:** MLRP / Architecture, Research & DevelopmentGHB Landscape A/S **Photographer:** Stamers Kontor **Client:** City of Copenhagen (CAU)

项目地点：丹麦，哥本哈根 建筑面积：140平方米 设计师：MLRP建筑研发事务所、GHB景观事务所 摄影师：斯达莫尔斯·康托尔 委托人：哥本哈根市政府（CAU）

Danish-American based architects MLRP has transformed an existing graffiti-plagued playground structure to an inviting and reflective pavilion as part of the new Interactive Playground Project in Copenhagen.

Funhouse mirrors are mounted on the gabled ends of this playground pavilion in Copenhagen, as well as behind the doors. This engages a play with perspective, reflection and tranformation. Instead of a typical closed gable façade, the mirrored gables creates a sympathetic transition between built and landscape and reflects the surrounding park, playground and activity.

Windows and doors are integrated in the wood-clad façade behind façade shutters with varied bent mirror panel effects.

At night the shutters are closed making the building anonymous. During the day the building opens up, attracting the children who enjoy seeing themselves transformed in all directions.

With simple means it has succeeded to transform an existing, sad and anonymous building to a unique and respectful installation in the newly renovated park.

The roof and façade is clad with heat-modified wood and the gables and shutters are clad with mirror polished stainless steel.

MLRP建筑事务所将一个深受涂鸦困扰的运动场改造成为深受欢迎的倒映馆，作为哥本哈根互动运动场项目的一部分。

奇幻屋镜子被安装在建筑两端的山墙和门上，在透视、反射和变形之间形成了有趣的互动。建筑没有采用典型的山墙外立面，镜面山墙在建筑和景观之间形成了对称的过渡，倒映出周边的公园、运动场和活动。

门窗嵌入木包层表面，设在百叶窗结构之后，拥有各种各样的曲面镜效果。

夜晚，闭合的百叶窗让建筑变得普通。白天，开放的建筑会吸引孩子们来从各个角度看到自己扭曲的影像。

项目利用简单的方式，成功地将一座不起眼的无名建筑改造成为公园里独特而受人欢迎的全新结构。

建筑屋顶和外立面由热改良的木材包裹，而山墙和百叶窗则覆有镜面抛光不锈钢。

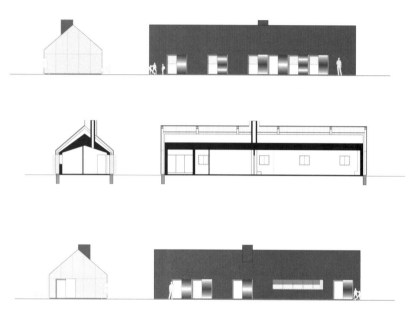

1. Funhouse mirrors are mounted on the gabled ends of this playground pavilion, as well as behind the doors.
2. MLRP has transformed an existing graffiti-plagued playground structure to an inviting and reflective building as part of the new Interactive Playground Project in Copenhagen.
3. The pavilion is clad in charred timber but its polished steel ends reflect the surrounding playground and trees.
4. During the day the building opens up, attracting the children who enjoy seeing themselves transformed in all directions.

1. 奇幻屋镜子被安装在建筑两端的山墙上和门后
2. MLRP建筑事务所将一个深受涂鸦困扰的运动场改造成为深受欢迎的倒映馆，作为哥本哈根互动运动场项目的一部分
3. 建筑包裹在烧焦的木材之中，抛光的两端则倒映出四周的操场和树木
4. 白天，开放的建筑会吸引孩子们来从各个角度看到自己扭曲的影像

5.6. At night the shutters are closed making the building anonymous.
7. Both convex and concave mirrors are mounted onto the backs of doors, which swing open when the building is in use to create an outdoor hall of mirrors.
8. Windows and doors are integrated in the wood-clad façade behind façade shutters with varied bent mirror panel effects.

5、6. 夜晚，闭合的百叶窗让建筑变得普通
7. 门后安装着凸透镜和凹透镜，当大门打开时，建筑就形成了一座独特的露天镜子大厅
8. 安装在木包层外立面上的门窗配有各种各样扭曲的镜面效果

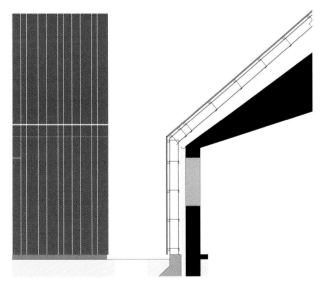
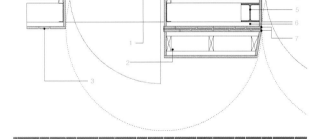
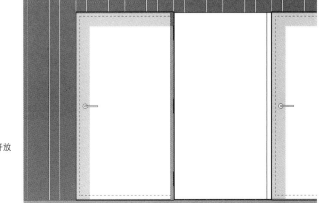

1. Door with new panel cladding	1. 新面板包层门
2. New façade panel – open	2. 新外立面板——开放
3. New wood cladding	3. 新木包层
4. New isolation	4. 新隔热层
5. New steel construction	5. 新钢结构
6. Under construction	6. 建设中
7. New wood cladding	7. 新木包层

VanDusen Botanical Garden Visitor Centre

范杜森植物园游客中心

Completion time: 2011 **Location:** Vancouver, Canada **Area:** 1,765 sqm **Designer:** Perkins+Will Canada
Photographer: Nic Lehoux **Client:** Vancouver Board of Parks and Recreation

竣工时间：2011年 项目地点：加拿大，温哥华 项目面积：1,765平方米 设计师：珀金斯+威尔 摄影师：尼克·勒乌
委托人：温哥华公园和娱乐设施董事会

VanDusen Botanical Garden's new Visitor Centre is designed to create a harmonious balance between architecture and landscape, from a visual and ecological perspective. Through mapping and analysing the Garden's ecology, the project team was able to integrate natural and human systems, restoring biodiversity and ecological balance to the site. The green roof and surrounding landscape were carefully designed to include native plants, forming a series of distinct ecological zones; a vegetated land ramp was included to connect the roof to the ground plane, encouraging use by local fauna; and old-growth trees were carefully preserved, facilitating an ecologically balanced system of wetlands, rain gardens and streams.

Inspired by the organic forms of a native orchid, the Visitor Centre is organised into undulating green roof 'petals' that float above rammed earth and concrete walls. Mimicking natural systems, the building is designed to harvest sunlight, store energy until needed, and collect water. Located on the Garden's prominent southeast corner, the 1,765-square-metre Visitor Centre transforms the entrance to heighten public awareness of the Garden and the importance of nature. With solid walls that protect visitors from the busy street and transparent walls that open the building toward the Garden, the Visitor Centre houses a café, library, volunteer facilities, garden shop, offices and flexible classroom spaces.

Designed to exceed LEED Platinum, the Visitor Centre is pursuing the Living Building Challenge – the most stringent measurement of sustainability in the built environment. Placing enormous constraints on projects, such as restricting the use of Red List Materials, only three projects worldwide have earned full certification. The Challenge also pushes buildings to achieve ground-breaking results: the Visitor Centre uses on-site, renewable sources – geothermal boreholes, solar photovoltaic, solar hot water tubes – to achieve net-zero energy on an annual basis. Wood is the primary building material, sequestering enough carbon to achieve carbon neutrality. Rainwater is filtered and used for the building's greywater requirements; 100% of blackwater is treated by an on-site bioreactor and released into a new feature percolation field and garden. Natural ventilation is assisted by a solar chimney, composed of an operable glazed oculus and an aluminium heatsink, which converts the sun's rays to convection energy. Summer sun shines on darker surfaces to enhance ventilation further. Located in the centre of the atrium, and exactly at the centre of all the building's various radiating geometry, the solar chimney highlights the role of sustainability by form and function.

Comprised entirely of FSC-certified Douglas fir, the panelised roof structure is composed of more than 50 different pre-fabricated roof panels that include electrical, sprinkler and audio-visual systems, which were integrated in the shop. Because of the necessity to precisely develop the complex curve geometry, the team designed the roof with Rhino software and physical models generated with rapid prototype three-dimensional printing.

1. The Visitor Centre was inspired by the organic forms of a native orchid.
2. A prominent bridge at the entrance is made of recovered fir from a former walkway on site.
3. The Visitor Centre is organised into undulating green roof 'petals'.

1. 游客中心从兰花的有机造型中获取了灵感
2. 入口桥梁的铺面来自于从场地上回收的铺路松木
3. 游客中心围绕着起伏的绿色"花瓣"屋顶展开

从视觉和生态角度来看，范杜森植物园的新游客中心在建筑和景观之间建立了和谐的平衡。通过对植物园生态的了解和分析，项目团队整合了自然与人类系统，恢复了场地的生物多样性和生态平衡。绿色屋顶和周边景观精心挑选了原生植物，形成了一系列独特的生态区；生长植物的坡地将屋顶和地面连接起来，大量采用了本地植物；成熟的树木得到了保护，促进了湿地、雨水花园和溪流之间的生态平衡。

游客中心从本地兰花的有机造型中获得了灵感，围绕着起伏的绿色"花瓣"屋顶展开。屋顶下方是素土夯实和混凝土墙壁。建筑模仿了自然系统，能够收获阳光、储存所需能源并收集雨水。这座1,765平方米的游客中心位于植物园的东南角，突出了植物园的形象和自然的重要性。密封的墙壁将游客和繁忙的街道隔开，而透明的墙壁则让建筑面朝植物园开放。游客中心内设有咖啡厅、图书馆、志愿者设施、花园书店、办公室和灵活的教室空间。

游客中心的设计超越了绿色建筑白金认证，追求"生态建筑挑战"——最严格的可持续建筑标准。项目面临相当多的限制条件，例如全部采用红色材料，全世界只有三个项目获得了全部认证。"挑战"还促使建筑实现开创性的成果：游客中心使用当地的可再生资源——地热钻孔、太阳能光电板、太阳能热水管——来实现年度零能源消耗。木材是主要的建筑材料，隔绝了足够的碳来实现碳中和。雨水的过滤和使用均符合建筑与水处理标准；而100%的黑水都经过就地生活反应器处理，被释放到过滤场地和花园。自然通风由太阳能烟囱辅助。烟囱由可控式玻璃圆孔和铝制散热器组成，将太阳辐射转化为对流能量。夏季，照射在深色表面的阳光能够促进通风。太阳能烟囱坐落在中庭的正中，也是建筑的几何正中，在造型和功能上都凸显了可持续设计的重要性。

屋顶结构全部采用森林管理委员会认证的花旗松，由50多块不同的预制屋顶板组成。屋顶上设有电气、洒水和视听系统，与商店相互融合。由于复杂的曲线造型需要十分精确，设计团队与犀牛软件公司共同设计了屋顶，利用实物模型快速生成了三维图纸。

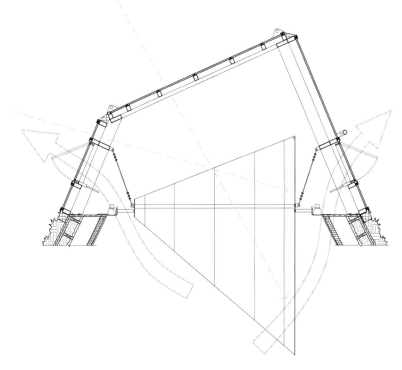

4. The Visitor Centre is designed to create a harmonious balance between architecture and landscape.
5. Green roof petals float above rammed earth and concrete walls.

4. 游客中心的设计在建筑和景观之间实现了平衡
5. 绿色屋顶画板飘浮在素土夯实和混凝土墙壁之上

6. The Visitor Centre transforms the entrance to heighten public awareness of the Garden.
7. The roof includes more than 50 different pre-fabricated roof panels composed of unique curved glulam beams.
8. Local, sustainably harvested wood was chosen as the primary building material.
9. Windows open automatically above the metal heat sink as part of the passive ventilation strategy.
10. The panelised roof structure is comprised entirely of FSC-certified Douglas fir wood.

6. 游客中心将入口变形,以突出公众对花园的认知
7. 屋顶由50多块不同的屋顶板组成,形成了独特的弧形胶合梁
8. 采用可持续砍伐的本地木材被选作主要的建筑材料
9. 金属散热器上方的窗口会自动开放,是被动式通风策略的一部分
10. 板块式屋顶结构全部采用森林管理委员会认证的花旗松

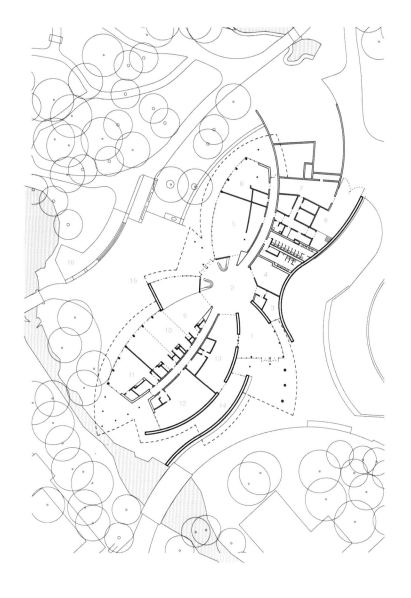

1. Arrival hall	9. Great hall	1. 到达大厅	9. 大厅
2. Atrium	10. Flex	2. 中庭	10. 灵活空间
3. Office	11. Classroom	3. 办公室	11. 教室
4. Intrep. Centre	12. Library	4. 说明中心	12. 图书室
5. Food service	13. Garden shop	5. 餐饮服务	13. 花园商店
6. Volunteer	14. Outdoor shop	6. 志愿者	14. 露天商店
7. Services	15. Livingston plaza	7. 服务区	15. 利文斯顿广场
8. Loading bay	16. Livingston lake dock	8. 装货间	16. 利文斯顿湖码头

Thomas Jefferson Visitor Centre and Smith History Centre at Monticello

托马斯·杰弗逊游客中心和史密斯历史中心

Completion date: 2007 **Location:** Charlottesville, USA **Area:** 5.26 ha **Designer:** Michael Vergason Landscape Architects, Ltd. **Photography:** Glenn Neighbors, Mike Merriam **Client:** Thomas Jefferson Foundation, Inc.

竣工时间：2007年 项目地点：美国，夏洛茨维尔 项目面积：5.26公顷 设计师：迈克尔·维尔贾森景观建筑事务所 摄影师：格伦·内博斯、麦克·梅里安 委托人：托马斯·杰弗森基金会

The Thomas Jefferson Visitor Centre and Smith History Centre project exemplifies a commitment to sustainable site development while creating a new landscape that is a quiet foil to Jefferson's exquisite 18th century mountaintop home and gardens. By limiting new site disturbance, the project protects critical views, archaeological resources, and habitat, simultaneously creating an educational, inspiring, and vibrant gateway to this UNESCO World Heritage site.

The project site includes a primary courtyard, greenroof meadow, terraces, landscaped paths, vistas, parking, and landscaped stormwater facilities, as well as the LEED Gold Thomas Jefferson Visitor Centre and Smith History Centre. The building complex consists of five pavilions: ticket centre, orientation theatre, interpretive galleries, a café, and gift/museum shop (including an outdoor garden sales area). The pavilions are organised around a central planted courtyard, beneath which the pavilions unite in a single floor plate providing staff and support space. The Courtyard, an outdoor lobby of sorts, serves a democratic function allowing guests to select the sequence of activities during their visit. The building's three levels follow the slope of the hillside and provide direct connections between interior space, the surrounding landscape and views to the forest beyond. The stacked building programme enables the pavilions to hug the topography with minimal profiles so as not to disrupt views from the top of the mountain, or overwhelm the scale of the historic home beyond.

The visitor's succession through the Piedmont Virginia forest before arriving at the Mountaintop is critical to the experience of Thomas Jefferson's Monticello. The landscape architect determined it was imperative for the project to maintain the intimate woodland experience of one's arrival to the place as a counterpoint to the expansive vistas and cultivated landscapes of the historic Mountaintop. As such, the landscape immediately surrounding the Visitor and History Centre plays an important role in preparing the visitor for their arrival to the site's historic core.

After evaluation of several potential building sites on the Monticello property, the project team ultimately chose to re-develop the previous visitor arrival facility parcel. The decision to build on the already developed land enabled re-use of significant existing infrastructure, including parking, vehicular circulation, and some stormwater facilities and utilities. A series of slope analyses were performed to situate buildings and additional parking facilities.

The Thomas Jefferson Visitor Centre and Smith History Centre welcomes guests as they begin their experience of Jefferson's quintessential plantation home and landscape, and helps send them away with a full, satisfying, and intellectually engaging visitor experience. The landscape and buildings of the Visitor Centre provide a place of calm and quiet consideration, respite, and education before and after visiting the infamous mountaintop home. By laying lightly on the land, this new complex illustrates primary principles of sustainable design and pays homage to Jefferson's ideals and commitment to innovation and thoughtful landscape design.

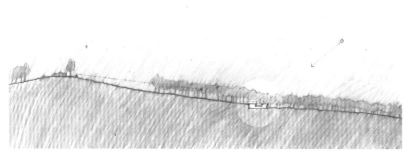

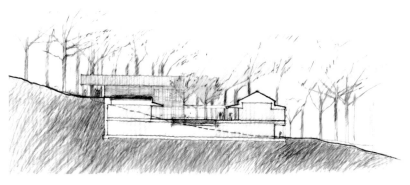

1. Staircase   1. 楼梯
2. Path to entrance   2. 入口小路
3. Fountain   3. 喷泉
4. Entrance   4. 入口

托马斯·杰弗逊游客中心和史密斯历史中心项目展示了可持续场地开发，为杰弗逊精致的18世纪山顶住宅和花园打造了一个全新的景观。通过限制新的场地干扰，项目保护了关键的景色、具有考古学价值的资源和生态栖息地，同时为这个世界文化遗产地打造了一个具有教育意义而又充满生机的门户。

项目场地包含主庭院、绿色屋顶草坪、露台、景观小路、远景、停车场以及景观暴雨保护设施组成，还有绿色建筑协会金奖认证的托马斯·杰弗逊游客中心和史密斯历史中心。建筑综合体由五个场馆组成：票务中心、定向剧场、说明展馆、咖啡馆和礼品商店（包含一个露天花园售货区）。场馆围绕着中央庭院展开，在地下一层统一起来，提供员工和辅助空间。作为一个露天大厅，庭院让游客可以在此自由选择想要参观的地点。建筑的三层楼沿着山坡地势而建，直接连接了室内空间、周边景观和远处的森林景色。层叠的建筑结构让场馆以最小的剖面与景观详解，既不会破坏山顶的景色，也不会削减后方历史住宅的气势。

游客穿过皮德蒙特弗吉尼亚森林到达山顶，这是参观杰弗逊庄园的重要体验。景观建筑师认为项目必须保留私密的树林体验，使其与辽阔的远景和精致的山顶庄园形成对比。同样的，游客和历史中心周边的景观对游客进入庄园前的准备也至关重要。

在评估了若干个潜在建筑场地之后，项目团队最终选择重新开发原有的游客到达设施。经过开发的场地上有大量的可再利用设施，如停车场、车道以及一些暴雨保护设施。他们对如何设置建筑和附加的停车设施进行了一系列分析。

托马斯·杰弗森游客中心和史密斯历史中心迎接着游客，为他们开始体验杰弗逊庄园和景观提供了准备，帮助他们完成更满意的参观体验。在进入著名的山顶住宅之前，游客中心的景观和建筑为游客提供了一个宁静思考、休息和获得信息的场所。通过贴近地面，这座新设施展示了可持续设计的主要原则，向杰弗逊的理念表达了敬意，同时也表现了建筑师的创造力和精心设计的景观。

Award: 2009 Maryland & Potomac Chapters of ASLA: Honour Award

获奖情况：2009年美国景观建筑师协会马里兰和波托马克分会荣誉奖

5. The visitor centre is organised around the central courtyard.
6,7. View to Courtyard

5. 场馆围绕着中央庭院展开
6、7. 庭院

7

1. The Greensward connects the visitor centre to the slave cemetery.
2. Slave cemetery
3. Central courtyard that organises the five pavilions, serving as an outdoor lobby
4. Ticket Pavilion
5. Orientation
6. Café
7. Exhibition
8. Retail
9. Shuttle stop
10. Bus parking
11. Car parking

1. 草皮将游客中心与奴隶公墓联系起来
2. 奴隶公墓
3. 中央庭院围绕着五个馆亭展开，是一个露天大厅
4. 售票亭
5. 方位介绍
6. 咖啡厅
7. 展览厅
8. 零售店
9. 公交车站
10. 公交停车场
11. 汽车停车场

8. Gallery with wood structure
9. Opening Day
10. The building follows the slope of the hillside and provides direct connections between interior space, the surrounding landscape and views to the forest beyond.
11. Green roof from above

8. 木质结构的长廊
9. 开放日
10. 建筑依地势而建，连接了室内空间、周边景观和远处的森林景色
11. 俯瞰绿色屋顶

内布拉方舟 Arche Nebra

Completion date: 2007 **Location:** Wangen, Germany **Designer:** Holzer Kobler Architekturen **Client:** Burgenlandkreis, Kreisverwaltung

竣工时间：2007年 项目地点：德国，万格 设计师：霍尔泽·柯布勒建筑事务所 委托人：布尔根兰县政府

In 1999, unlicensed treasure hunters unearthed a remarkable archaeological relic: a 3,600-year-old sky disc made of bronze inlaid with gold. It depicts complex constellations and the symbol of the solar barge representing the sun's nightly passage from west to east. An architectural competition was launched to design a public archaeological centre and an observation tower that would showcase the disc and come to symbolise the region.

The desigenrs chose the symbol of the solar barge for the construction of the centre. Visible from far away, the body of the building is covered with yellow anodized aluminium and appears to float above the glass-encased entry level, in which the admission desk and café are located. The 60-metre-long abstracted ship houses two exhibition rooms and the planetarium.

The open vertical atrium connecting the ground floor to the first floor symbolises the relationship to the heavens. The rough plastered base housing the seminar rooms and offices appears to emerge from the hillside.

The permanent exhibition explores the site of the discovery and the historical environment, while the immense picture window surrounding it presents visitors with a vista of the Mittelberg mountain and the observation tower. The panoramic window of the temporary exhibition space offers a view of the Unstrut river.

The exact site where the disc was discovered is marked by a 30-metre-high conical tower, creating a landmark that can be seen from far around. Widening towards the top, inclined 10 degrees to the north and divided by a vertical crevice extending over its full height to mark the sommer solstice, it replicates the function of the sky disc as a solar calendar. Once a day the sun passes through the vertical opening, indicating the line of sight towards the Brocken mountain some 80 kilometres away, just as the Brocken served as a reference point for the sky disc in the Bronze Age.

1999年，未受许可的寻宝者们发掘出一件非凡的考古文物：拥有3,600年历史的铜制镀金天盘。它描绘出复杂的星座图像，并以太阳船象征着太阳由西向东的夜间轨迹。当地政府举办了一次建筑竞赛，要求设计一个公共考古中心和瞭望塔，以便展示天盘，也为该区域提供一座标志性建筑。

设计师选择了太阳船的象征来建造中心。从远处看，建筑主体上覆盖着一层黄色电镀铝，仿佛悬浮于玻璃围住的入口层（内设有行政前台和咖啡厅）之上。60米长的抽象大船内设置着两间展览厅和星象仪。

开放的垂直中庭将一楼和二楼连接起来，象征着地面与天堂的关系。粗糙的灰泥地下层内设有研讨室和办公室，看似从山坡上浮现。

永久展览区探索了场地的历史发现和历史环境，四周宽阔的观景窗为参观者呈现出米特伯格山和瞭望塔的远景。临时展览空间的全景玻璃窗则展现了温斯特鲁特河的景色。

天盘确切位置由一个30米高的圆锥形高塔标志出来，打造了从远处都十分注目的地标。瞭望塔向上逐渐扩宽，朝北倾斜10度，一条垂直的裂缝贯穿全塔，标志出冬至/夏至日。它再现了天盘作为阳历计时器的功能。太阳每天穿过裂缝一次，标志出朝向80公里外布罗肯山的视线，正如布罗肯山是天盘在青铜时代的参照物一样。

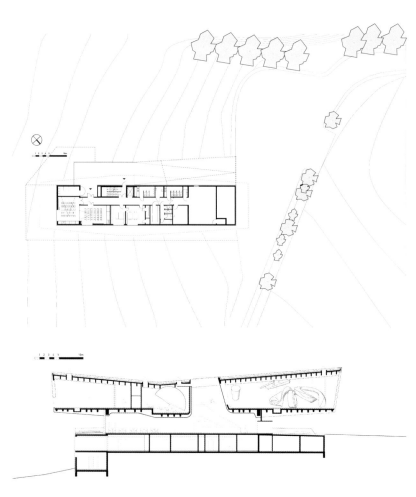

1. An overlook to the architecture and landscape
2. The panoramic window offers a view of the Unstrut River.

1. 远眺建筑和景观
2. 全景玻璃窗展现了温斯特鲁特河的景色

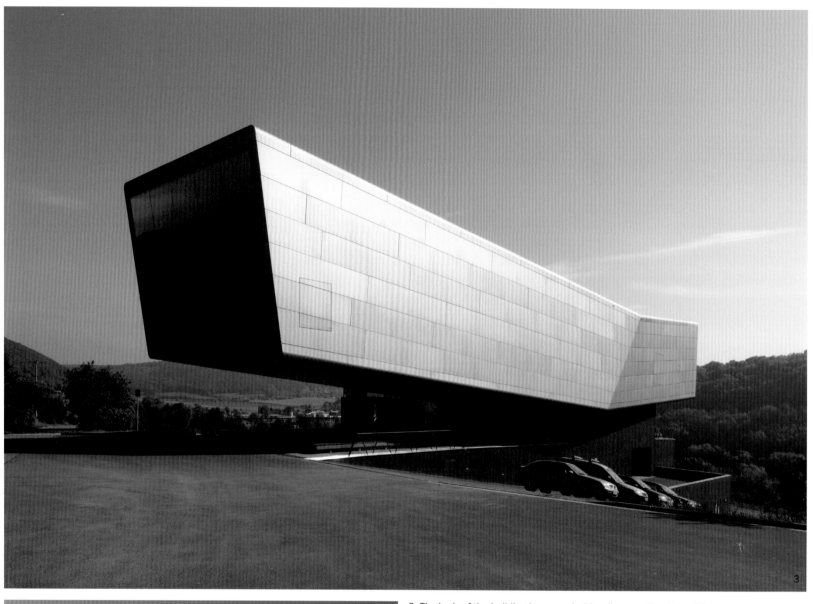

3. The body of the building is covered with yellow anodised aluminium.
4. The conical tower widens towards the top and inclines 10 degrees to the north.
5. Entrance rest area under the stairs

3. 覆盖黄色电镀铝的建筑
4. 瞭望塔向上逐渐扩宽,朝北倾斜10度
5. 楼梯下的入口休息区

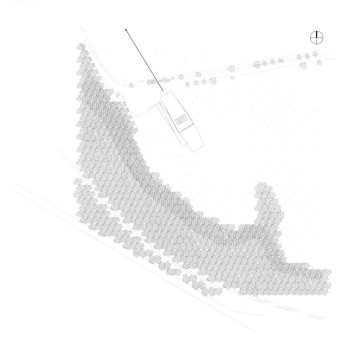

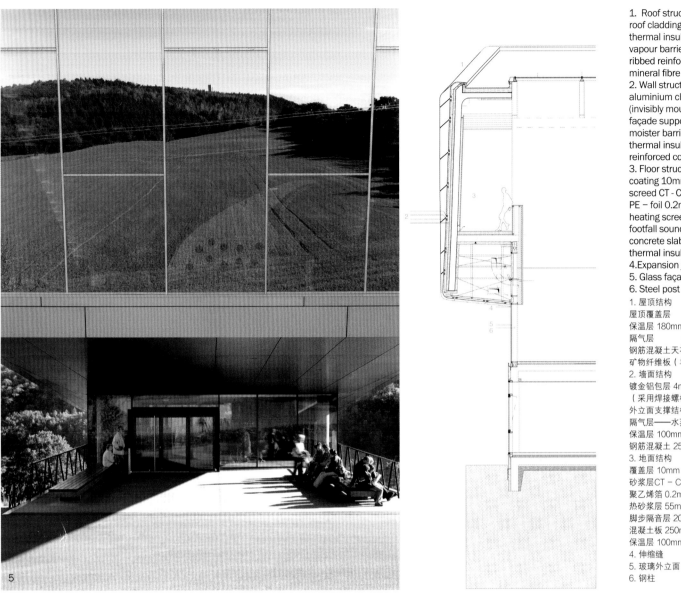

1. Roof structure:
roof cladding
thermal insulation 180mm
vapour barrier
ribbed reinforced concrete ceiling 200mm
mineral fibre boards (fleece-covered) 80mm
2. Wall structure:
aluminium cladding anodized in gold 4mm
(invisibly mounted with welding studs)
façade support structure – aluminium
moister barrier – vapour permeable
thermal insulation 100mm
reinforced concrete 250mm
3. Floor structure:
coating 10mm
screed CT - C30 - F5 75mm
PE – foil 0.2mm
heating screed 55mm
footfall sound insulation 20mm
concrete slab 250mm
thermal insulation 100mm
4. Expansion joint
5. Glass façade
6. Steel post

1. 屋顶结构
屋顶覆盖层
保温层 180mm
隔气层
钢筋混凝土天花板 200mm
矿物纤维板（羊毛覆盖）80mm
2. 墙面结构
镀金铝包层 4mm
（采用焊接螺柱隐形安装）
外立面支撑结构——铝
隔气层——水蒸气可渗透
保温层 100mm
钢筋混凝土 250mm
3. 地面结构
覆盖层 10mm
砂浆层CT – C30 – F5 75mm
聚乙烯箔 0.2mm
热砂浆层 55mm
脚步隔音层 20mm
混凝土板 250mm
保温层 100mm
4. 伸缩缝
5. 玻璃外立面
6. 钢柱

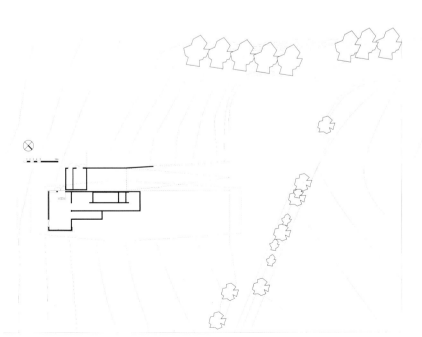

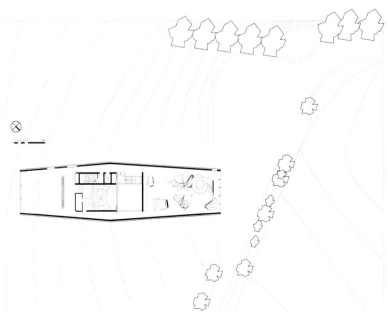

Bai Sha Wan Beach and Visitor Centre

白沙湾海水浴场旅客服务中心

Location: Taipei, China **Completion date:** 2007 **Site area:** 37,651 sqm **Construction area:** 5,657 sqm
Designer: Wang Weijen Architecture + North Reign Space Design **Client:** North Coast and Buddha Mountain Scenic Spot Administrative Ofce of Taiwan Tourism Bureau

项目地点：中国，台北 竣工时间：2007 占地面积：37,651平方米 建筑面积：5,657平方米 设计师：王维仁建筑设计研究室＋北域空间设计事务所 委托人：台湾交通部观光局北海岸及观音山风景区管理处

The design of Bai Sha Wan Beach and Visitor Centre reflects strong wind in winter, sand dune and strong sunlight in summer. Coordinating with topography and landscape, the architecture tries to change the traditional binary relation between people, landscape and architecture, expecting the architecture to become a part of the whole landscape such as trees and rocks.

The visitor centre is transformed and expanded from an old architecture of 1980s. The design connects the existing building to the new visitor centre through the floor with gentle slope. Meanwhile, visitors can walk up to the grass slope on the roof through the topography of white sand. The L shape folded volume clings to the contour line and retaining wall, creating a plaza facing the sea. The green roof and folded plates interact with the interior activities, move of the light and shadow and the coastal landscape. The green roof's arrangement of up and down not only coordinates with the undulating of the topography and grass slope, but also takes advantage of the folded plates with various sizes, bringing in natural light and ocean view. Simultaneously, the massive roof creates an extensive, natural-ventilated porch and an semi-open space, providing large shade area for visitors. From the parking lot behind the architecture, visitors pass through the entrance porch, walk along the functional paths of exhibition, service, F&B, showering to the beach. They would become an integral part of the nature and enjoy the beauty of ocean and sky.

As part of Taiwan landscape transformation of tourism plan, the design of Bai Sha Wan Beach and Visitor Centre wishes to improve the surrounding environment of Bai Sha Wan Beach and strengthen the service facilities through the transformation of the old architecture, making Bai Sha Wan an important node in north coast tourism line. The existing office building of Bai Sha Wan aged terribly, suffering from leakage and inadequate facilities, urging for rebuilt or renovated. Although the surroundings are developing in a mess, they get a potential for transformation thanks to the flow of people brought by the coach. The plan also proposes to integrate the other potential tourist resources, including the continuous white sand shore, the rock coast to the east and Linshanbi Peninsula to the west. It needs a complete plan and integral architectural strategy to improve the landscape environment to drive the future reservation and development of Bai Sha Wan.

In the condition of limited budget, the whole plan defines the key issues and focuses on two directions of development: firstly, in the term of landscape, the designers add a plank walkway and many seats along the east, expand the visitors' activities from sea bathing to the surrounding rock coast and grass slope, take full advantage of the rich coastal topography and flora landscape, and provide various beach experiences for visitors. Secondly, on the term of architecture, the designers reserve the existing structures and available facilities, improve the exterior waterproof and appearance, re-plan the internal functions, extend the architectural volume, and add more exhibition area, F&B, shower and changing facilities. Besides, they plan to build a small relaxing facility on the grass slope of the rock coast for the visitors to the rock area.

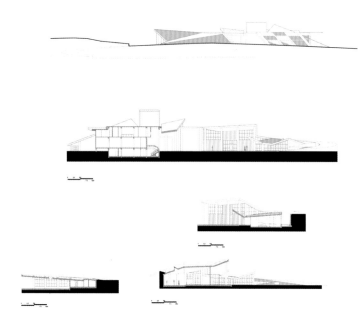

1. An overview of the site   1.基地外景
2. Slanted skylights   2.斜顶天窗
3. Lawn plaza   3.草地广场

白沙湾海水浴场旅客中心的设计，反映了冬日的强风、沙丘与夏日强烈的日照。建筑配合地形与地景，企图改变人与地景及建筑的传统二元关系，期望建筑就像大树、岩石一样成为地景的整体。

这是从一栋20世纪80年代的老旧建筑改造和加建而成的。设计以缓坡楼板将原有建筑与新建的旅客服务中心连接，并使游客能顺着白沙地形走上建筑顶的草坡。L形的曲折量体紧贴着等高线与挡土墙，形成一个面海的大广场。屋顶草坡与折板则反映了室内的活动、阳光的阴影移动与海岸的景观。覆草的屋顶升起下降的安排，除了配合地形与屋顶草坡的起伏，也利用向北侧开起的大小折板，带入自然光线与海景。同时，大块的屋顶更是为了营造大面积、自然通风的穿廊与半户外空间，提供大量遮阳空间供游客活动。观者由草坡建筑背侧的停车场，穿过框景的入口穿廊，沿着建筑一侧的展示、服务、餐饮、冲洗等功能路步向沙滩，体验海天一色、天人合一的意境。

作为2004年政府策划的台湾地景改造套装旅游系列之一，白沙湾海水浴场旅客中心的设计，希望透过对北观处旧建筑的改造，改善白沙湾海水浴场的周边环境，扩充旅客中心的服务设施，使白沙湾成为台湾北海岸旅游系列的重要节点。在规划议题上，原有的白沙湾北观处办公楼由于建筑老旧，漏水严重加上设施不足，亟待重建或改善。临近的聚落周边环境虽然发展凌乱，却也因为海水浴场带来的人流而产生转型的潜力。规划也提出了整合白沙湾附近其他具潜力的旅游资源，包括绵长的白沙海岸，以及东侧的礁石海岸与西侧的麟山鼻半岛等景观资源，亟待一个全面的规划构想与整体的建筑策略来改善地景环境，以带动白沙湾未来的发展。

整个规划设计在极有限的经费下，界定了关键的议题，并将设计集中往两个方向发展：第一，在景观上，沿着沙滩由旅客服务中心向东增建木栈道平台与座椅，将游客的活动由海水浴场区延伸到周边的礁石海岸与草坡，利用白沙湾丰富的海岸地形与植物景观，提供旅客更多样性的海岸经验；第二，在建筑上，保留原有建筑结构与可用的设施，改善外墙防水与外部造型，重新规划内部使用功能，并扩大延伸建筑量体，增加展示空间，以及餐饮与淋浴更衣等功能。规划上同时计划在礁石海岸草坡增建一个小型休憩设施，作为礁石区游客的据点。

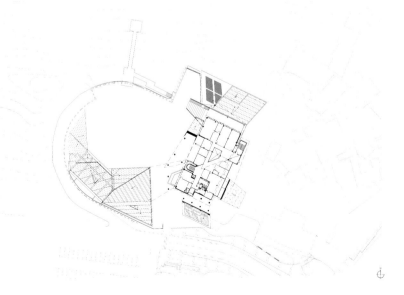

4. Grass slope on the roof
5. Restaurant terrace
6. Administrative entrance

4. 屋顶草坡
5. 餐厅平台
6. 行政入口

7-8. The elevation facing the sea
9. Interior exhibition space

7、8. 面海立面
9. 室内展示空间

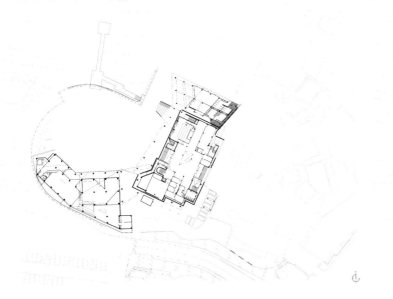

1. Entrance
2. Exhibition
3. Gallery
4. Visitor centre
5. Dressing room
6. Catering
7. Plaza
8. Bathing beach

1. 入口
2. 展示
3. 廊道
4. 旅客中心
5. 更衣
6. 餐饮
7. 广场
8. 海水浴场

Hanil Cement Visitors Centre and Guesthouse

韩一水泥游客中心和宾馆

Completion date: 2009 **Location:** Danyang, South Korea **Site area:** 3,957 sqm **Construction area:** 1,044 sqm **Designer:** BCHO Architects Associates **Photographer:** Yong Kwan Kim, Wooseop Hwang

竣工时间: 2009年 项目地点: 韩国, 丹阳 占地面积: 3,957平方米 建筑面积: 1,044平方米

The purpose of this project is to educate visitors about the potential for recycling concrete. In Korea concrete is the primary building material so it is imperative to begin to re-use the material as buildings are brought down. The Visitors Centre is a working example with alternative types of construction and landscaping using recycled and different casting techniques of concrete in new and interesting ways. Concrete has been broken into smaller pieces and recast in various materials creating both translucent and opaque tiles. The displays will continue to evolve and change at the Visitors Centre as new techniques are designed.

As the site is situated beneath a forested hill, the architect mainly took consideration of the access to the building, the direction in which it faced, the natural lighting, and its relationship with the Hanil headquarters building and factory. In addition, literal account of movements of spaces that influenced the open layout.

The site is located to the westernmost part of the factory, adjacent to Mt. Sobaek National Park. The existing land had been changed much to facilitate the movement of trucks to the cement factory. First of all, the archtects tried to restore the damaged original mountains and forest. In order to revive the landscape, they brought in earth to fill the courtyard between the two buildings, along the original mountain ridge. The seeming flow of the mountains from the west leads to the reception and dining in the inner courtyard of the building. In the in-between spaces they allowed people to experience obstructed space and non-obstructive spaces.

While following the linear placement and movement of land and earth, the architects came up with ideas for the new building façade. They cast fabric-formed concrete walls to the east façade, evoking images of the forest behind. There are four openings in the eastern wall and long vertical windows have been created in their in-between spaces. Through the windows, one can see how concrete is produced at the factory. Behind the two larger openings, one can see the courtyard of the Visitors' Centre and the cafeteria next to the courtyard, which is encircled by a water garden.

项目的目标是向参观者宣传再生混凝土的潜力。在韩国，混凝土是主要的建筑材料，因此，混凝土的再利用势在必行。游客中心是运用再生材料和新式混凝土预筑技术打造替代建筑和景观的鲜活范例。混凝土被碾成小块，在各种各样的材料中进行重筑，用以打造半透明和不透明的瓷砖。随着新技术的开发，游客中心会不断进行进化和改造。

由于项目场地位于茂密的山上，建筑师首先考虑了建筑的入口设计、朝向、自然采光以及建筑与韩一公司总部大楼和工厂的关系。此外，还有空间运功对开放式布局的影响。

场地坐落在工厂的最西端，紧邻小白山国家公园。原有的土地经过了多次改造，以方便水泥厂的卡车活动。首先，建筑师试图修复受损的山脉和森林。为了让景观重获新生，他们用土填充了山脉旁、两楼之间的庭院。山脉走势自西向东，直达建筑内院的前台接待处和餐厅。在中间的空地，人们可以尽情体验无障碍的景观和空间。

建筑师遵循了土地的线性布局和运动，提出了全新的建筑外立面设计方案。他们在东立面采用了纤维混凝土墙，突出了后方森林的景色。东墙上设有四个开口，其间设有细长的窗口。通过窗口，人们可以看到工厂中混凝土的制作流程。两个较大的开口让人可以观看到游客中心庭院和自助餐厅（餐厅紧邻庭院，四周环绕着水上花园）的景象。

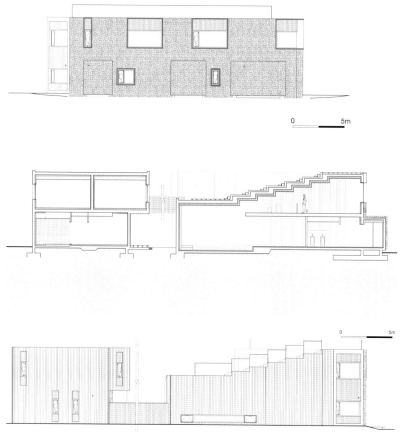

1. Guest house fabric formwork detail
2. Guest house entrance
3. Guest house east & south façade
4. Guest house east façade
5. Guest house entry courtyard
6. Guest house water garden
7. House entry hall
8. Recycled concret panels
9. Stairway

1. 宾馆纤维框架细部
2. 宾馆入口
3. 宾馆东南立面
4. 宾馆东立面
5. 宾馆入口庭院
6. 庭院水景
7. 入口门厅
8. 回收利用的水泥板
9. 楼梯

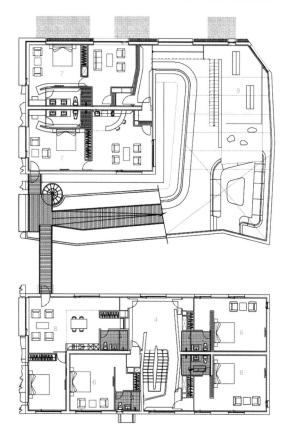
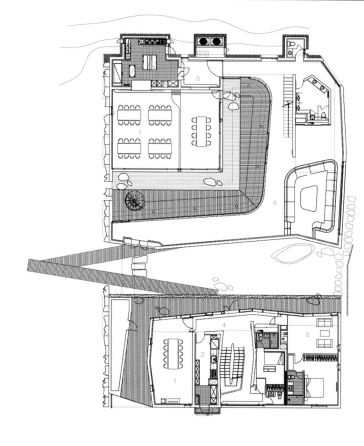

1. Dining
2. Kitchen
3. Reception
4. Hall
5. Suite
6. Guest room
7. VIP suite
8. Managers suite
9. Library

1. 餐饮区
2. 厨房
3. 接待处
4. 大厅
5. 套房
6. 客房
7. VIP套房
8. 经理套房
9. 图书室

Vanke Triple V Gallery

Completion date: 2011 **Location:** Tianjin, China **Site area:** 16,850 sqm **Building area:** 750 sqm
Designer: Ministry of Design **Photographer:** CI&A Photography, Edward Hendricks

竣工时间：2011年 项目地点：中国，天津 项目面积：16,850平方米 建筑面积：750平方米 设计师：MOD设计事务所 摄影师：CI&A摄影、爱德华·亨德里克斯

Designed as a permanent show gallery and tourist information centre for China's largest developer Vanke, MOD's dramatic design for the TRIPLE V GALLERY has become an icon along the Dong Jiang Bay coastline.

Despite its obvious sculptural qualities, the building's DNA evolved rationally from a careful analysis of key contextual & programmatic perimeters – resulting in the Triple V Gallery's triangulated floor plan as well as the three soaring edges that have come to define its form.

The client's programme called for three main spaces: a tourist information centre, a show gallery & a lounge for discussion. Requiring their own entrances, the tourist centre and the show gallery are orientated to separate existing pedestrian pathways and can be operated independently.

An extension of the show gallery, the lounge area is where discussions are conducted. This space takes advantage of the panoramic views of the coastline and comprises a sculptural bar counter.

Tectonically, the building responds to the coastal setting and is finished in weather-sensitive Corten steel panels on its exterior and timber strips on the interior walls and ceiling for a more natural feel.

MOD为中国最大的开发商——万科公司——设计了一座永久的展示画廊和游客信息中心，他们对3V画廊的戏剧性设计成为了东江湾的一座地标。

除了明显的雕塑特征之外，建筑的特征演变自对于主要周边环境的理性分析，最终形成了3V画廊的三角形楼面布局和3个高耸的屋檐。

委托人要求打造三个主要空间：游客信息中心、画廊以及休息讨论室。由于游客中心和画廊需要独立的入口，它们朝向独立的人行道，可以独自运营。

作为画廊的扩展部分，休息区是人们进行讨论的地方。这一空间利用了海岸线的全景，拥有一个造型独特的吧台。

建筑在造型上与海岸环境相互呼应，在外立面采用了柯尔顿钢板，室内则通过木条来呈现更加自然的感觉。

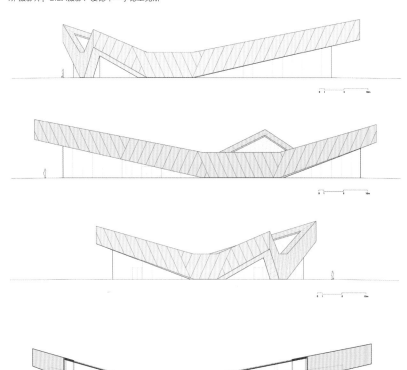

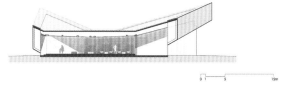

1. A night view of the building
2. Tectonically, the building responds to the coastal setting.
3. Triple V Gallery's triangulated floor plan and the 3 soaring edges

1. 建筑夜景
2. 建筑在造型上与海岸环境相互呼应
3. 三角形楼面布局和3个高耸的屋檐

4

4. The building is finished in weather-sensitive Corten steel panels on its exterior.
5. The unique sculptural bar counter can enjoy the panoramic views of the coastline.
6. Panoramic window along the bar

4. 外立面采用了柯尔顿钢板
5. 造型独特的吧台可以欣赏海岸线的全景
6. 吧台全景窗

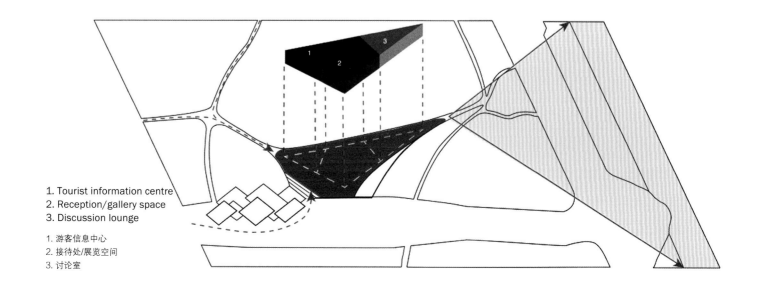

1. Tourist information centre
2. Reception/gallery space
3. Discussion lounge

1. 游客信息中心
2. 接待处/展览空间
3. 讨论室

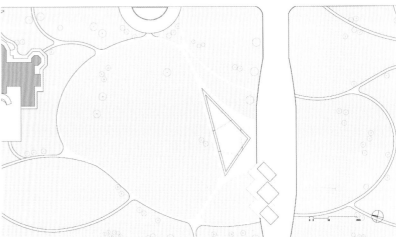

7. Timber strips on the interior walls express a natural feel.
8. Triangle lighting belts echo with the architectural form.
9. Lounge area

7. 室内墙壁采用木条来呈现自然的感觉
8. 三角形的灯带与建筑外形相呼应
9. 休息讨论区

Diagram
场地图解

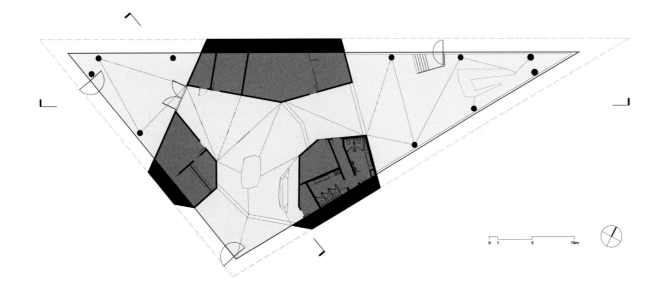

9

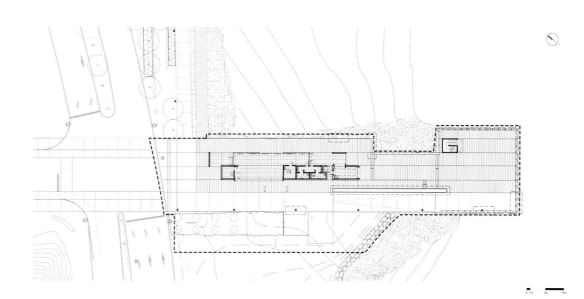

斯纳菲尔斯托法游客中心 Snaefellsstofa, Visitor Centre

Completion date: 2010 **Location:** Skriduklaustur í Fljótsdal, Iceland **Area:** 750 sqm **Designer:** ARKÍS ARCHITECTS **Photographer:** Sigurgeir Sigurjónsson, Birgir Teitsson, Arnar ór Jónsson **Client:** Vatnajökulsjógarur – Vatnajökull glacier

竣工时间：2010年 项目地点：冰岛，斯奇杜卡劳斯托尔 项目面积：750平方米 设计师：阿尔奇斯建筑事务所 摄影师：西格吉尔·西格琼森、比格尔·泰特森、阿尔纳尔·强森 委托人：瓦特纳冰川公园

Visitors are intended to walk up to the building, much like climbers progressing onto the glacier. Thereby, visitors effectively experience the glacier's grandeur. The building stands alone and automobiles and other vehicles are kept to the side and their visual impact is reduced by screening them off with vegetation and landscaping walls. Facilities are provided for visitors to enjoy the site's exterior spaces. Pedestrian paths, ramps and decks are built from local wood and their forms steer the approach of visitors up to the building. Staff facilities such as workshop, garage and technical rooms are located so that they are not visible from the approach.

Material palettes are clear and simple. Staff and service spaces are the foundation for the operations. The 'Rock Foundation' is a cast in place concrete structure with wood board texture both inside and out, with insulation in between the concrete layers. In addition, larch paneling is intertwined into the concrete walls both inside and out.

The exhibition and education axis, the Ice Stream, is constructed of wood. For that reason, that part of the building will make creaking noises depending on the weather, much like a glacier. The axis is clad in locked dark-brown copper on the exterior, but with wood on the inside. Crossing the two axes is the information path, connecting the axes together.

Spaces can easily be combined and connected, both indoor and outdoor. For example, meeting room and central hall can be opened up to the external amphitheater. The same goes for café and exterior deck. The design of exterior spaces has focused especially on accessibility and universal design.

Concept and inspiration for interior and exterior lighting design is sought in the ever-changing light reflection of the glacier, its eternal creative force and how ice mediates light. Not unlike being inside an ice cave, where the irregular placements of liner lights in the ceilings reflect the light qualities inside glacial fissures and circular recesses in the ceilings are reminiscent of icicles as are vertical exterior lights. In all cases, the lighting concept is based in the creation of defused light.

The lighting of various spaces is designed in accordance with the ideas described above, but the spaces have different lighting concepts depending on the use and how the design can fulfill requirements for exhibition design, ergonomics and energy use. Indoor and outdoor lighting is informed by the building's location and is designed to minimise light pollution, while maximising the use of daylight.

The visitor centre has been designed with special emphasis sustainable design and is undergoing BREEAM assessment. Visits to the centre can thereby have positive impact on environmental experiences for guests and communicate the message of sustainable design out to the community about the importance of the environmental issues in architecture.

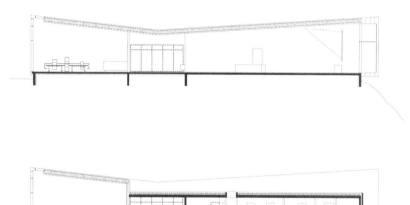

Section a
剖面a

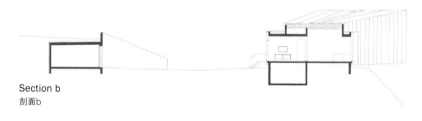

Section b
剖面b

1. Approach
2. The marriage between the building and the landscape
3. Turf roof

1. 入口通道
2. 建筑与景观的结合
3. 草皮屋顶

1

游客必须走上建筑,就像登上冰川一样,以此体验冰山的宏伟。建筑"遗世独立",机动车和其他交通工具都停放在别处,被植被和景观墙隔开。建筑的建造目的是为游客提供享受场地露天空间的设施。人行道、坡道和平台都由本地木材建成,吸引着游客走向建筑。工坊、车库和技术室等员工设施都设在游客看不见的地方。

建筑在材料选择上清晰而简单。员工和服务空间是运营的基础。"岩石地基"是现场浇筑的混凝土结构,内外全部采用木板纹理,混凝土层之间设有隔热层。此外,落叶松板在内外的混凝土墙面上错综复杂。

展览和教学中轴——"冰流"由木材构成。因此,建筑的一部分会随着气候的变化而发出吱吱嘎嘎的声音,就像冰川一样。中轴外部由深棕色铜板包裹,内部则是木材。穿过两条中轴的是信息走道,它将二者连接在一起。

无论是室内还是室外,空间的结合和连接都十分简单。例如:会议室和中央大厅都通往露天剧场。咖啡厅也直通露天平台。户外空间的设计特别聚焦于通路和统一的设计。

室内外灯光设计的概念和灵感来自于冰山不断变换的反光,它永恒的创造力量以及冰对光线的传导。与在冰洞里一样,天花板上衬灯的不规则布局反映了冰川裂缝内的光线特点,而天花板上像冰锥一样的圆形槽则是垂直的外部灯。

各个空间的灯光设计与上文的描述相同,但是根据其功能的不同,各个空间的灯光拥有不同的主题,满足了展览设计、人体工程学和能源使用上各种要求。

游客中心的设计特别注重可持续性,正在环境评估测试。游客中心将会对游客的环境体验起到积极的影响,向社区传递可持续设计的信息以及建筑中环境问题的重要性。

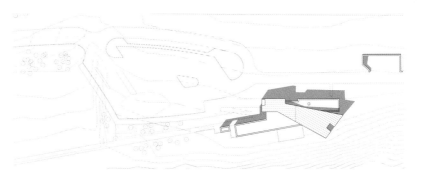

4. Connection path from Gunnar's-house
5. Connection of material, larch and concrete
6. South side view

4. 贡纳之屋的连接走道
5. 材料、落叶松和混凝土的联系
6. 建筑南面

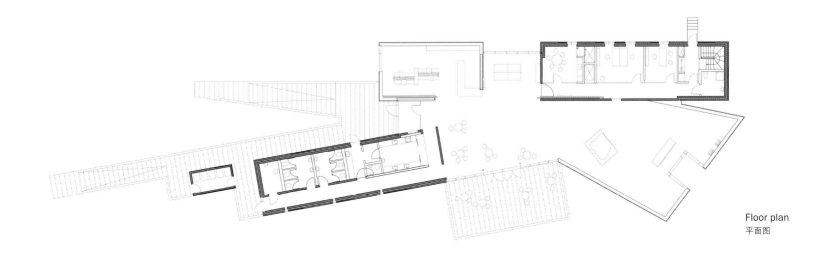

Floor plan
平面图

New Visitor Centre of the Pannonhalma Arch-abbey

帕农哈尔玛大修道院游客中心

Completion date: 2010 **Location:** Pannonhalma, Hungary **Gross storey area:** 2,727 sqm **Building area:** 446 sqm **Designer:** Roeleveld-Sikkes Architects **Photographer:** Tamás Bujnovszky, Hans Molenkamp, Balázs Kátai **Client:** Benedictine Arch-abbey of Pannonhalma

竣工时间：2010年 项目地点：匈牙利，帕农哈尔玛 楼层面积：2,727平方米 建设面积：446平方米 设计师：罗利维尔德-斯琪斯建筑事务所 摄影师：塔马斯·布吉诺夫斯基、汉斯·莫伦坎普、巴拉斯·卡泰

In 2006 a tender had been issued for the design and construction of the new visitor centre of the Benedictine Pannonhalma Arch-abbey to improve its tourism, infrastructure and gastronomic services. The purpose of this development is to meet the demand of the continuously growing number of visitors.

The competition tender has been awarded to Roeleveld-Sikkes Architects who not only focused on the design of the visitor centre but also paid close attention to the surrounding buildings to ensure that it fits into the architectural history of the environment.

The development site is located on the Kosaras Hill in the immediate vicinity of the World Heritage listed Arch-abbey. Prior to the development, the Kosaras Hill used to be covered with asphalt and serve as a parking lot for the tourists who visited the Arch-abbey without available quality infrastructure or hospitality services. Therefore the most important aspect of the architectural concept – its first fundamental architectural principle – was to restore the natural environment of this site.

The reception building of the Arch-abbey was designed by György Skardelli in 2003 and has a strong visual dominance and a symbolic role as a 'gatehouse', and forms the connection between the established and the contemporary architecture in this historical environment. The new visitor centre of Roeleveld-Sikkes Architects takes into consideration the significant architectural impact of the existing reception building. Its position and shape – the second fundamental architectural principle – were determined by its relation to the reception building.

The visitor centre is a two-storey building but the lower floor is hidden by the hill therefore only the transparent glass façade of the upper floor is visible from a distance. This innovation ensures that the building is well proportioned to both the hill and the reception building, and well suited to the environment. The rehabilitation of the Kosaras Hill – the restoration of its natural environment – covers the parking area, the bicycle and luggage stores, which are hidden from view as you arrive and are also concealed when viewed from either the Arch-abbey or from the valley. The visitor centre is surrounded by a one-way road, which leads the tourists' cars and buses to this backside area. The conference hall on the lower floor has a capacity for 120 guests, which gives the opportunity to host a wide variety of events.

The service areas are on the lower floor to enhance the transparent and flexible utilisation of the upper floor: the a la carte and the buffet restaurant are floating above the hill overlooking the valley.

These improvements allow visitors to spend their time pleasantly until the commencement of the exhibition tour in the Abbey. Furthermore, the visit can be expanded into a day-long family trip by visiting the renovated arboretum, the new herb garden and the pilgrimage house as well.

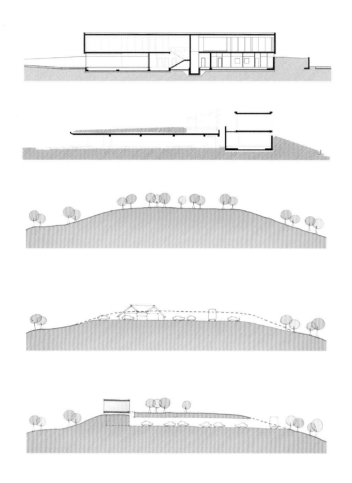

1. The lower floor is hidden by the hill therefore only the transparent glass façade of the upper floor is visible from a distance.
2. The project is located on the Kosaras Hill in the immediate vicinity of the World Heritage listed Arch-abbey.
3. Visitor centre and lawn

1. 从远处只能看到上层的透明玻璃墙面，底层被山体隐藏
2. 项目位于科萨拉斯山上，紧邻世界文化遗产大修道院
3. 游客中心和草坪

4

5

2006年，帕农哈尔玛大修道院决定建造一个新的游客中心，以提升它的旅游业、基础设施和餐饮服务，并对此进行了竞标。本次开发的目的是满足日益增长的游客数量需求。

竞标最终花落罗利维尔德-斯琪斯建筑事务所，他们不仅注重游客中心的设计，还对周边建筑投入了足够的重视，以保证它能够融入周边环境的建筑历史。

开发场地位于科萨拉斯山上，紧邻世界文化遗产大修道院。在开发之前，科萨拉斯山上覆满了柏油，是大修道院的游客停车场，上面没有任何良好的基础设施或游客服务。因此，建筑概念最重要的一部分——也是它的首要建筑原则——就是修复场地的自然环境。

大修道院的接待楼由戈伊尔吉于2003年设计，具有强烈的视觉效果，是象征性的"门楼"。它将历史环境中的已有建筑和现代建筑连接了起来。罗利维尔德-斯琪斯建筑事务所设计的新游客中心详细考虑了原有接待楼的建筑影响力。它的位置和造型——也是第二重要的建筑原则——都由其与接待楼的关系决定。

游客中心高两层，底层被山体隐藏起来，从远处只能看到上层的透明玻璃墙面。这种创新保证了建筑与山脉和接待楼之间的良好比例，同时也与环境十分契合。科萨拉斯山自然环境的修复覆盖了停车场、自行车和行李仓库，它们都被巧妙地隐藏了起来，游客在入口乃至大修道院和其他山谷都看不见。游客中心四周的一条单向道路引领游客的汽车和公交到达这个后方区域。底层的会议厅可容纳120名客人，适合举办各种各样的活动。

服务区设在建筑底层，提升了二楼的通透感和灵活性；点餐餐厅和自助餐厅设在楼上，俯瞰着山谷的风景。

这些改进让游客能够在参观大修道院之前有一个愉快的体验。此外，作为家庭一日游，游客们还可以参观翻修的植物园、新建的草本园和朝圣屋。

Award:
ALUTA (Association of aluminium windows and façades) – The best designer(s) award for creating an architectural design of aluminium-glass structure

获奖情况：
铝窗和外墙协会——铝-玻璃结构建筑最佳设计师奖

4. Hidden entrance to the visitor centre
5. A night view of the visitor centre
6. Outdoor rest area
7. Terraced architecture and landscape

4. 隐藏的游客中心入口
5. 游客中心夜景
6. 户外休息区
7. 高低变换的建筑与环境

8. Small plaza under the sunshade
9. Stairs to the first floor
10. The a la carte and the buffet restaurant are floating above the hill overlooking the valley.

8. 遮阳伞下的小广场
9. 通向2楼的楼梯
10. 餐厅和自主餐设在楼上,俯瞰着山谷的风景

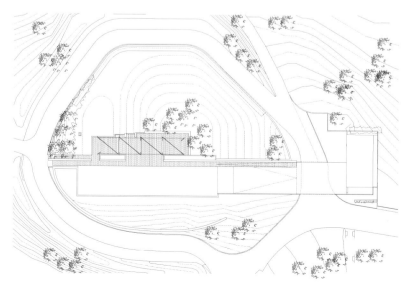

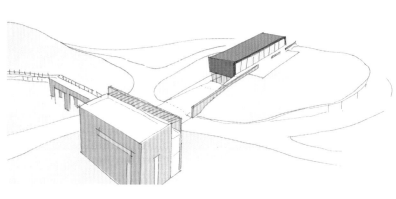

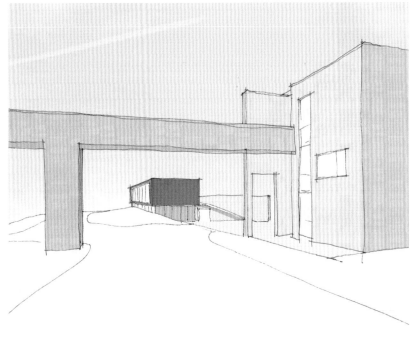

1. Entrance   1. 入口
2. Reception   2. 接待处
3. Dining   3. 就餐区
4. Staircase   4. 楼梯

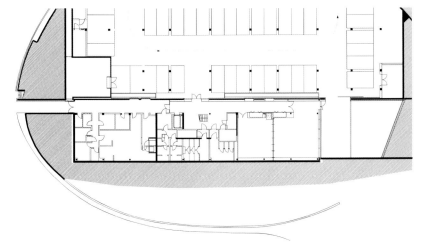

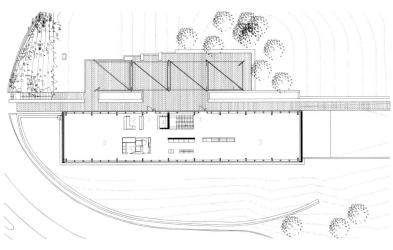

Cabrera Islands National Park Visitors Centre

卡布雷拉岛国家公园游客中心

Completion date: 2010 **Location:** Cabrera Islands, Spain **Total area:** 4,526 sqm **Construction rea:** 3,343 sqm **Designer:** Estudio de Arquitectura Alvaro Planchuelo **Photography:** Estudio de Arquitectura Alvaro Planchuelo

竣工时间：2010年 项目地点：西班牙，卡布雷拉岛 总面积：4,526平方米 建筑面积：3,343平方米 设计师：阿尔瓦罗·普兰奇洛建筑工作室 摄影师：阿尔瓦罗·普兰奇洛建筑工作室

Situated at the beginning or the end of an important area of natural interest, the site is made up from the remains of old Mediterranean pine forests. The proximity of the beach, the port and the buildings isolated between the pine forests suggests the construction of a small building, using the rest of the site as a park and placing the main part of the project under the slope, protected from the heat.

The proposed distribution of the rooms, routes and communication means that the visitor can make a virtual journey through space, time, culture and the natural values of the National Park and, by extension, the Western Mediterranean. The model used will be an isolated island 'cliff', similar to many existing in the archipelago. The buried part of the 'cliff' will show the undersea world. The exterior will show the terrestrial environment and the interior, the Mediterranean culture.

The trajectory will begin in the undersea area, treated like a continuous journey from the depths to the surface. The interior decoration of the aquariums continues through the visitors' gallery, giving the impression of diving to the bottom of the sea. Two panoramic lifts submerged in an aquarium will serve as a link with the fresh air of the headland, on the first floor where the terrestrial environment of the park will be interpreted. From a viewpoint on the uppermost part there will be a descending ramp displaying a mural of Mediterranean culture. The journey ends on the ground floor, crossing an open-air water park with sculptures representing Mediterranean islands. Architecture, sculpture, art and nature unite in one building to explain the secrets of the Mediterranean.

项目场地位于自然保护区的一端，遍布了古老的地中海杉木森林。场地紧邻海滩、港口以及由树林隔开的建筑，因此要求建造一座小型建筑，将场地的剩余部分作为公园。项目的主要部分设在坡下，以免受热气侵袭。

空间、路线和交流的分配意味着游客能够通过空间、时间、文化和国家公园的自然价值进行一次虚拟旅行，乃至了解西地中海地区。设计以群岛地区常见的孤岛悬崖为模型。悬崖埋在地下的那部分将被用于展示海底世界。建筑外观将展示陆地环境，而室内则将展示地中海文化。

旅程从海底开始，由深至浅。室内装饰的的水族缸沿着游客的走廊而延续，让人仿佛置身海底。两个潜入水下的全景电梯连接了二楼的新鲜空气，将人带到陆地环境。建筑的最高处有一条下降的坡道，展示了以地中海文化为主题的壁画。参观旅程终结于一楼，人们穿过一个露天水上雕塑公园（象征着地中海的岛屿）走出游客中心。建筑、雕塑、艺术和自然在一座楼内融为一体，解释了地中海的秘密。

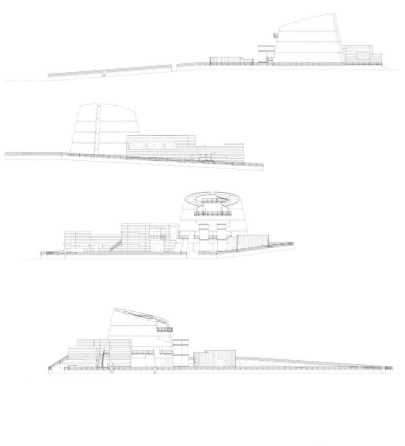

1. Entrance to the visitor centre
2. A night view of the architecture
3. Outdoor water sculpture garden
4. Deck path to the information centre

1. 游客中心入口
2. 建筑夜景
3. 露天水上雕塑公园
4. 通向信息中心的甲板路

5. Interesting ramp entrance
6. Visitor centre and sculpture garden
7. Service centre
8. Path paved with cobbles

5. 有趣的通道式入口
6. 游客中心和雕塑公园
7. 服务中心
8. 卵石铺成的小路

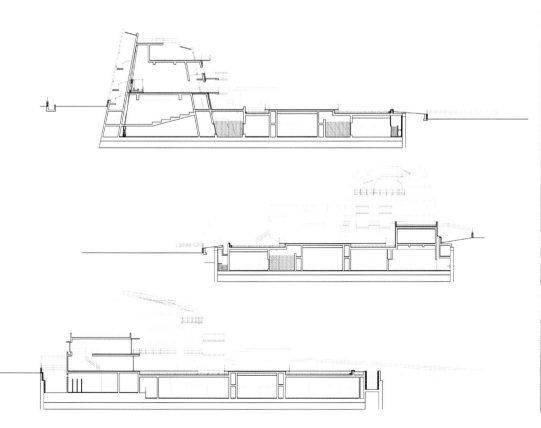

9. The paths symbolised Mediterranean islands lead to the visitor centre.
10. Landscape terrace
11. Gift shop

9. 象征着地中海的岛屿的路通向游客中心
10. 观景台
11. 纪念品商店

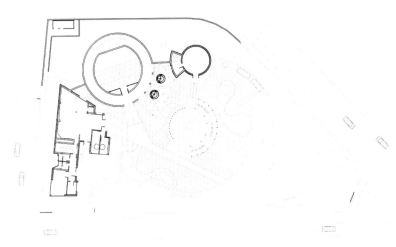

Ground floor plan
一层平面图

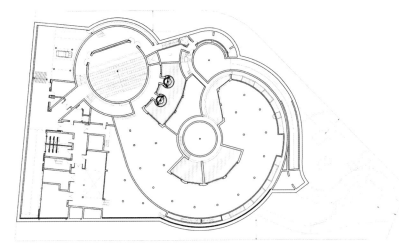

Basement plan
地下室平面图

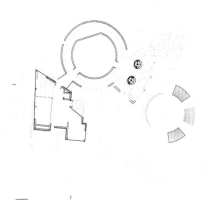

First floor plan
二层平面图

Second floor plan
三层平面图

Roof plan
屋顶平面图

12

13

12. The interior journey of underwater world
13. The interior decoration of the aquariums continues through the visitors' gallery.
14. Staircase
15. A mural of Mediterranean culture
16. Rounded glass skylight

12. 仿佛置身海底的室内旅程
13. 水族缸沿着游客的走廊而延续
14. 楼梯
15. 展示了以地中海文化为主题的壁画
16. 圆形玻璃天窗

生活艺术公园 PAV – Park of Living Art

Completion date: 2008 **Location:** Torino, Italy **Area:** 24,000 sqm **Designer:** Gianluca Cosmacini
Photographer: Mattia Boero, Valentina Bonomonte, Leo Gilardi

竣工时间：2008年 项目地点：意大利，托里诺 项目面积：24,000平方米 设计师：赞布罗塔·科斯玛西尼 摄影师：马蒂亚·布罗、瓦伦蒂娜·波诺蒙特、里奥·格拉尔迪

The PAV, Park of Living Art is located in a former industrial area of Turin, where, up until the early 90s of the last century, metal and mechanical industries produced automobile components. Today this transformation has become newly incorporated into the city's productive mechanisms. Where once thousands of car suspension springs were made, today the production of value is sustained by new protagonists and by new processes, such as that shown in the publicity of Turin' Environmental Health Agency, with the image of a banana peel sticking out of a jewellery box, shown in ads on the city buses.

During this last transformation, the park's territory was abandoned for many years, a vague terrain, a place suspended in time where soon, or so it was said in the neighbourhood, an equipped green area would be created.

In this terrain in transit, wild in its own way, utilised as a dumping ground for construction débris, in this Third Landscape, a concept elaborated by Gilles Clément, if observing carefully, one can be witness to a pioneer life animated by the settlement of many species of vegetation. A transitory space endowed with its own vitality. A space projected toward an unprecedented transformation, becoming the matter of accordance among the many protagonists, in order to accomplish the completion of the Park of Living Art as a substitute for the foreseen equipped park area.

The idea of the plans for the Park of Living Art has been specified in this phase of the process and by these observations: a park, not totally organised in a formal language, that is structured and defined by the process of living things – plants, animals, people and through the relationships these form with each other and which generate unexpected transformations, not determined a priority by the urban landscape. The Park's planning takes into account the requests and relationships with the artists, their relationships with the environment, with its constraints and potentialities, and with the public and its social commitments, in a system of relations where the place of the art works' production is the same as that of their exposition, and the place of the exposition is the territory where the works, relating to one another, contribute to the construction process of this fragment of landscape. Thus, the Park of Living Art proposes itself, according to a definition by Piero Gilardi, as 'an uninterrupted construction site, an interweaving dialogue of experience open to innovative alternatives, in homologation with the living systems of the biosphere'.

In short, to use a concept expressed by Nicolas Bourriaud, the Park of Living Art can be meant as 'a place of negotiation between man and things'. Bourriaud introduces his idea of a Park of Living Art in the flux of this elaboration, an idea of a contemporary art park, as an artistic territory undergoing development and not a container, satisfied with just hosting projects and exhibitions, but rather, a model of sustainable, lasting development between artistic practices and the exposition space that produces and displays them.

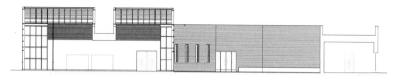

1.2. Detail of the building green roof with the artwork
3. North view of the pedestrian square, the building hill and the green outside areas
4. South view of the building hill, Trèfle and the green outside areas feature

1、2. 绿色屋顶和艺术品的细部设计
3. 步行广场、建筑和外层绿地区域的北侧景色
4. 建筑山、特莱弗里和外层绿地的南侧景色

3

4

生活艺术公园位于都灵的前工业区，直到20世纪90年代，那里的金属和机械工厂都在制造汽车元件。现在，这次改造已经成功地融入了城市的生产机制。从前制造成千上万个汽车悬簧的地方，如今已经在生产价值上通过新主角和新工艺得到了传承。正如都灵环境健康署的宣传——城市公交上的广告上画着香蕉皮从珠宝盒里露出来。

经过最后的改造，被废弃多年的公园领域、一个模糊的领地、一个命运悬而未决的场所，被设计成了美妙的绿地。

在这片已经成为建筑废料垃圾场的场地上，在这个由吉勒斯·克莱蒙特所提出的"第三景观"上，多种植物的定居让这里充满了生气。这个临时的空间赋予了自身活力。这个空间正面临史无前例的改造，集多种功能于一身，以便实现生活艺术公园的完工，使其成为完整公园区域的替代品。

生活艺术公园的概念来自于以下元素：这座正式的公园由鲜活事物——植物、动物、人的生活过程以及他们之间的相互联系所构建，形成了意想不到的转变，并不受城市景观先天条件所影响。公园的规划考虑了艺术家的要求和关系，他们与环境的关系，对公众和社会做出了承诺。艺术作品的生产和展示同时进行，而相互联系的艺术品在展示空间里对景观建造流程起到了很大的作用。因此，根据皮耶罗·杰拉迪的说法，生活艺术公园将自身打造成"一个连续的建筑场地，与创新经验形成了交织的对话，与生物圈的生态系统形成了同化反应"。

简而言之，在采用尼古拉斯·布里沃所表达的概念中，生活艺术公园意味着"一个人与事物的协商之地"。布里沃在精心设计的公园中引入了自己的观念，打造了一座现代艺术公园，使其成为经历艺术发展的领地，而不是一个容器。这不仅是一个项目和展览，而是一个可持续模型，展现了艺术实践与生产和展示艺术实践空间的持续开发过程。

5. The park's main entrance through the greenhouse and the pedestrian square
6. Night view of the inner court
7. Inside view of the greenhouse
8. Interior entrance

5. 通过花房和步行广场进入公园主入口
6. 内庭夜景
7. 花房内部
8. 室内入口

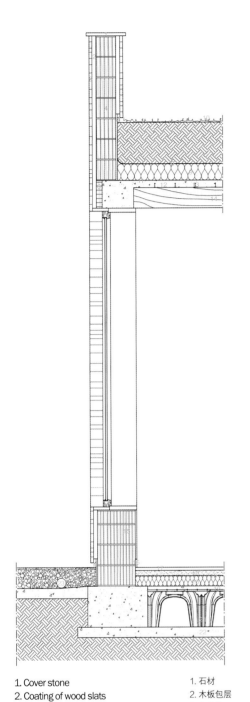

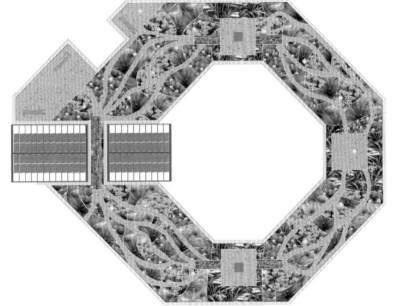

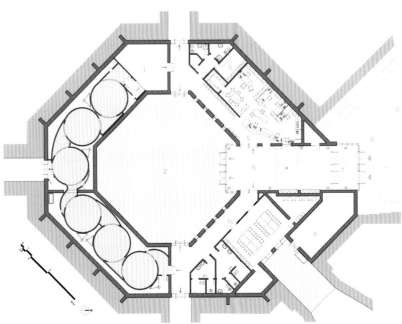

1. Cover stone	1. 石材
2. Coating of wood slats	2. 木板包层
3. Wood uprights	3. 木支柱
4. Masonry parapet	4. 石栏杆
5. Metal flashing	5. 金属盖片
6. Brick	6. 砖块
7. Soil by cultivation	7. 培养土
8. Non-woven fabric	8. 无纺布
9. Bituminous waterproofing in two layers	9. 两层沥青防水层
10. Panels	10. 面板
11. Coating of lacquer bitumen cold	11. 冷沥青覆盖层
12. Cast concrete with additional mesh	12. 浇筑混凝土，附有网眼
13. Pin connectors	13. 接头连接
14. Wooden floor boards in adjacent spruce	14. 云杉木地板
15. Brick blocks	15. 砖砌结构
16. Skirting stone	16. 壁脚石
17. Linoleum	17. 油布
18. Background in light-based conglomerate of expanded clay	18. 膨胀黏土砾石背景
19. EG sheet	19. EG板
20. Cellular glass panels	20. 泡沫玻璃板
21. Crawl space	21. 管线空间
22. Drainage pipe	22. 排污管
23. River pebbles	23. 鹅卵石

1. Exhibition space	1. 展览空间
2. Court	2. 庭院
3. Local technical	3. 本地技术区
4. Offices	4. 办公室
5. Reception	5. 接待区
6. Greenhouse	6. 温室
7. Entry hall	7. 入口大厅
8. Pedestrian plaza	8. 步行广场
9. Technical centre	9. 技术中心
10. Polivelente room	10. 波利维伦特厅

戈罗波维卡亭 Pavilion in Grebovka

Completion date: 2010 **Location:** Prague, Czech Republic **Construction area:** 370 sqm **Total area:** 3,100 sqm **Designer:** SGL Projekt s.r.o **Photographer:** Filip lapal

竣工时间：2010年 项目地点：捷克，布拉格 建筑面积：370平方米 总面积：3,100平方米 设计师：SGL项目公司 摄影师：菲利普·斯拉帕尔

The pavilion designed by architect Josef Schulz was built in the beginning of 1890s. It is an integral, urban, architectural and artistic part of the Gröbe Villa (designed by Antonín Barvitius) and Gröbovka Park, which represents the Late neo-Romantic work of garden architecture.

The building no. 2188 which served as a shooting range and skittle alley, is a garden triple-wing single floor pavilion built northwards of the villa. It has a light seasonal construction. The original load-bearing walls are half-timbered nogged with facing bricks.

The building had been used for its original purpose only shortly. Later, the municipality adapted it to a nursery. This children facility had been recently in operation (after numerous building adaptations – lowering of ceilings, removal of a glazed wall with a door in a changing room, gas installation, construction of a shelter with a terrace, establishment of a heating system with a boiler room and oil storage, etc.). Due to the fact that the pavilion was rebuilt to a nursery and used all year round, it was exposed to considerable temperature differences and increased wood moisture, particularly in winter. The reason for this moisture was condensation of vapours on the building envelope which has not adequate thermal resistance.

Besides rot, wooden joists, roof truss and wall timbers are damaged by feeding of larvae of wood-destroying insect.

The main aim is to recover the historical pavilion and open it for the public, to use it and its vicinity as a place of relaxation and leisure, to establish sanitary facilities for park visitors and to restore the authentic historical building including substantially damaged art-and-craft elements – carved decoration, wall timbering, paintings and to restore them to maximum extent. It is considered to implement already totally lost elements preserved only in drawings – skittle alley, wooden dadoing, and entrance doorways.

The recovery scheme of the pavilion is the 'restoration of artistic work'. The project plans to maintain and recover original appearance of the historical building and, thus, does not comply intentionally with current standards for development and returns the building to the maximum extent to its original appearance from the year 1888.

In response to the transparent expression of the historical building, the complementary building is covered by a fully glazed envelope along its entire parameter. The visible structural system combines wood, steel and glass. In relation to historical half-timbered walls, the building tectonics is emphasised by contours of columns in the façade and steel tie rod bracing. An expressive roof landscape of the Schulz's pavilion is complemented with bracketed pergolas with wooden fins overhanging the roof of a complementary building and its vicinity. The termination of the west wing is reflected in a roof overlap and creation of covered terrace. The building is connected to the existing historical pavilion by means of glazed 'neck' with a ramp.

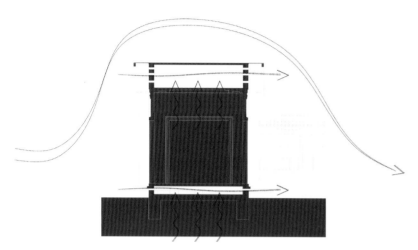

Air flow around the building reduces the humidity in the constructionthe under floor and roof ventilation had been restored surrounding terrain lowered by 0.7m to the year 1881 level

建筑周围的空气流动降低了施工现场的湿度，令地板下和屋顶的通风更顺畅，周围的地势较1881年降低了0.7米。

1. Restored historical building and the complementary building
2. The architecture is located in Gröbovka Park.
3. In response to the transparent expression of the historical building, the complementary building is covered by a fully glazed envelope along its entire parameter.

1. 修复的历史建筑和新建的附属楼
2. 建筑位于戈罗波维卡公园
3. 与历史建筑的通透感相呼应，附属楼四周全部采用了玻璃外立面

4

5

由建筑师约瑟夫·舒尔茨设计的亭馆建于19世纪90年代初期。它是一座完整的城市建筑,是戈罗伯别墅和戈罗波维卡公园的艺术元素,而公园是新浪漫主义园林的代表。

作为一座射击场和幢柱游戏球场,2188号建筑是一座三面单层花园建筑,位于别墅的北侧。它是一座季节性建筑,原始承重结构由木结构贴面砖组成。

建筑的原始功能只使用了很短的时间。不久,市政当局将它用作一家托儿所。这座儿童设施近期还在使用之中(建筑历经了大量改造:降低天花板、在更衣室以大门替代玻璃幕墙、煤气装置、露台庇护所、由锅炉房提供的供暖系统、油料仓库等)。由于建筑已被改造成托儿所,全年使用,它面临着可观的温度差和增加的木材湿度,特别是在冬天。这种潮湿源于建筑表面上水蒸气的凝结,而建筑外壳本身也没有足够的耐热性。

除了腐坏,木托梁、屋顶架和墙面木板还遭受到了害虫的破坏。

项目的主要目的是修复历史结构,使其面向公众,并以它作为一个休闲的场所,为公园游客提供卫生设施并真实地修复历史建筑。项目要最大程度地修复受损的工艺元素——刻花、木墙板绘画作品。项目将重现仅存于图纸中的元素——幢柱游戏球道、木刨槽和入口门廊。

建筑的修复计划是一项"艺术品修复"工作。项目计划保留并恢复历史建筑的原貌,并不完全遵从当前的开发标准,最大限度地将建筑恢复到1888年的面貌。

与历史建筑的通透感相呼应,附属楼四周全部采用了玻璃外立面。可见的结构系统由木材、钢和玻璃组成。与历史木墙相对应,外墙立柱的轮廓和钢拉杆支柱凸显了建筑的结构。舒尔茨所设计的建筑上方采用了广阔的屋顶景观,与现代建筑上方的木架绿廊遥相呼应。建筑的西翼在屋顶叠加和带顶平台中得到了反映。建筑与原有的历史建筑通过一个玻璃连廊连接起来。

4. The building is connected to the existing historical pavilion by means of glazed 'neck'.
5. The project returns the building to the maximum extent to its original appearance from the year 1888.
6. Restored historical building with transparency
7. Lawn in front of the building

4. 玻璃连廊连接新建筑与原有的历史建筑
5. 项目恢复历史建筑的原貌,最大限度地将建筑恢复到1888年的面貌
6. 恢复的具有通透感的历史建筑
7. 建筑前的草坪

8. Rest area in the glass corridor
9. Steel tie rod pillar
10. Old skittle lane

8. 玻璃连廊中的休息区
9. 钢拉杆支柱
10. 古老的幢柱游戏球道

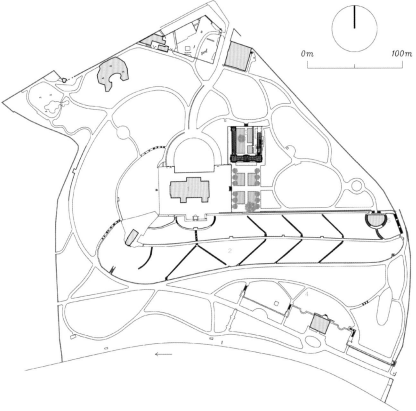

1. Villa
2. Winery yard
3. Pavilion
4. Lower landscape
5. Wine cellar

1. 别墅
2. 酿酒厂
3. 场馆
4. 下层景观
5. 酒窖

1. Inaccessible area of park
2. Building last used as a nursery
3. Inaccessible area of park
4. Rose garden
5. Café
6. Summer café
7. Skittle alley

1. 公园不可进入区
2. 被用作托儿所的建筑
3. 公园不可进入区
4. 玫瑰园
5. 咖啡厅
6. 夏日咖啡厅
7. 撞球场

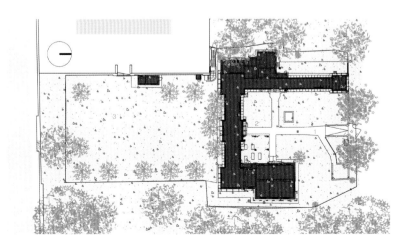
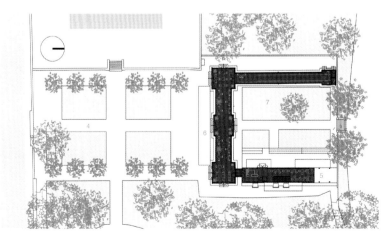

2010 Taipei International Flora Exposition

2010台北国际花卉博览会

Completion date: 2010 **Location:** Taipei, China **Area:** 1,830,119 sqm **Designer:** shu chang & associates architects **Photographer:** Zheng Jinming **Client:** Taipei City Government

竣工时间：2010年 项目地点：中国，台北 项目面积：1,830,119平方米 设计师：张枢建筑师事务所 摄影师：郑锦铭 委托人：台北市政府

The main design task of 2010 Taipei International Flora Exposition is not the independent pavilions, but the reestablishment of the existing football field, roads, park, landscape hills and water features, converting them into a unity suitable for large-scale park such as the flora exposition.

The project develops around exhibition and specifies the arrangement of pavilions. Since most of the Yuanshan Park are existing buildings, how to utilise the space, circulation and supporting facilities and unite various pavilions are the main task of architectural design.

Yuanshan Park contains two most important parts of the Expo: first is the gateway on the west, which is the only node connecting Taipei's MRT system (Red line Yuanshan Station); second is the international certificated main pavilion, EXPO Dome, located in the middle of Zhongshan Football field, which will hold a series of competitions and exhibitions of AIPH and is the key to the Expo's international acknowledgement.

In face of such a large-scale exhibition event like the flora exposition, with so many scattered sites and spaces of different qualities, the designers consider four significant design principles: Recognition, Circulation, Ecology and Environmental Protection.

Recognition has two meanings: one is the conspicuous visual sign which gives identity for the park; the other is the clear space sequence which gives the park easily-recognised orientation. The designers transform the flower forms into patterns and use them on the pavings and walls. Besides, the entrance pergola also uses flower as its form. Made of simple steel structure and colored steel panels, the pergola is foled into several pieces of petals. The steel beams are arranged from right angles to leaf vein shapes. The columns turn outward organically and are divided into different colour sections, avoiding the stiffness of basic steel structure. The designers hope these changes can connect the public and flowers at the entrance.

Circulation includes the transition between outside circulation and circulation inside the park, and the relationship between the internal circulation and the events. How to respond to the circulation of 8 million visitors in 6 months is a great challenge. Yuanshan Park is the west entrance of the Expo, which is the only node with MRT and the connection of two interchanges of the highway. Therefore, in fact it is the main gateway of the Expo.

The designers attempt to facilitate the visitors to use various public transportations, in order to reduce the pressure that the circulation puts on the neighbourhood and the roads, and to create a safe and accessible pedestrian space in the whole park.

On one hand, Ecology is related with the theme of the Expo. The Expo only lasts for half a year, but the environmental effect of Yuanshan Park is permanent. How to avoid fierce disturbance of the events in the 6 months, the recycle of the infrastructure and the recovery of environment are all important issues. In the park, the designers try their best to preserve every existing tree and only move several roadside trees at the entrance for the sake of the open space. The other important eco consideration is to set the public space semi-open, in order to fit Taipei's warm winter, mainly for the energy effectiveness of natural

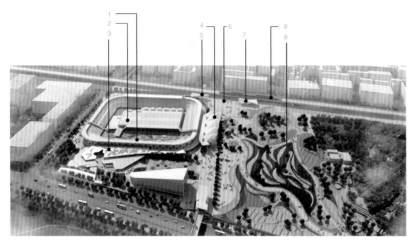

Zone Diagrams / 区位说明
1. Rainbows / 七道彩虹
2. EXPO Dome / 争艳馆
3. Snack seating area / 零食座位区
4. Ticket checking / 验票栅门（棚内）
5. Ticket booth / 售票亭
6. Gathering plaza / 集合广场
7. Ticket booth / 售票亭
8. Yuanshan MRT station / 圆山捷运站
9. Flower sea / 花海区

1. Sea of flowers in the flora exposition / 博览会圆山园区的花海
2. Entrance plaza / 入口广场
3. EXPO Dome and the rainbow / 争艳馆及七道彩虹

4

5

ventilation and natural light.
Environmental protection, energy conservation and carbon emission reduction are important public issues in recent years. Therefore, 2010 Flora Expo inevitably takes environmental protection as starting point of design. It develops in two terms: one is to save the construction cost; the other is the environmental protection of the materials. Since the Flora Expo is a temporary venue, the designers aim to build the largest space, the best circulation plan and a successful exhibition atmosphere in minimum budget.

2010台北国际花卉博览会圆山园区的主要设计课题不是独立的新展馆,而是重塑既有的足球场、道路、公园、山丘和行水区,转换为一个具整体性,而且适合花博这种大型活动的园区。

该项目以展示内容为主,指定了展场的位置,在圆山园区大多是既有建筑;因此,如何利用空间、动线和支援性设施将各展馆串连成一个整体,是建筑设计的任务。

圆山园区在整个花博中最重要的部分有二,一是西侧门户的地位,花博只有这个点与台北捷运系统相连(红线圆山站),二是国际认证花博的主展馆:争艳馆,在纲要计划中被配置在中山足球场中间,这个馆内将举行AIPH(花博的认证机构)所规定的一系列竞赛和展览,也是花博之所以可以被国际认证的关键。

面对花博如此大规模的展示活动,又涉及如此多分散和性质不同的场地和空间,设计师认为有四个重要的设计原则:辨认、动线、生态和环保。

"辨认"有双重含意,一个是容易辨认的视觉符号,赋予园区自明性,另一方面是清晰的空间序列,赋予园区易于辨识的方向感。设计师用花的造型转换为图案,不只用在铺面或墙面上,也是入口廊道棚架的造型,棚架的建材只是非常一般的钢构和彩色钢板,设计师把顶棚裂解翻折成数个如花瓣的片状,钢梁的排列从一般的直角改为如叶脉的形状,柱子较有机的向外倾斜,色彩分段,打破基本钢构的僵硬,这些变化希望能给民众在入口就可以和"花"联上关系。

"动线"包含外部动线和园区动线的转换,以及园区内动线和活动间的配合关系,如何应付6个月内800万人次的参观者的流动是一大挑战。圆山园区是花博的西门,是唯一有捷运的点,也是高速公路两个交流道之间的点,因此,事实上是花博的主门户。

设计师努力的方向是使各种公共运具都能方便的上下客,并降低该动线对邻近社区和道路的压力降低,以及塑造下车后安全无障碍的纯步行空间,再用步行贯穿整个圆山园区。

"生态"一方面和花博主旨有关,而且花博只有半年,圆山园区的环境却是永久的,如何使环境不受半年高强度活动太大的干扰,以及再利用花博的建设或恢复环境,都是重要课题。在圆山园区设计师努力保存每一株既有的树木,只在入口处不得已的移了几株路树,否则空间实在开阔不起来。另一个重要的生态考量是将公共空间设为半户外形式,配合台北的暖冬,主要还是为了自然通风采光的节能考量。

环保节能减碳为近年来一项重要的公共议题,2010花博不可避免也应该从环保角度作为设计出发点,其可由两大部分来谈,一为节省的营建成本;另一为营建材料本身的环保。因花博为临时性展场,我们的目标是以最少造价营造最大空间、最好的动线规划,并成功塑造展览园区气氛。

4. Entrance pergola
5. Night view of the entrance pergola, with columns turning outward organically
6. EXPO Dome is one of the most important parts of the flora expo

4. 入口长廊的棚架造型
5. 入口长廊夜景,柱子较有机的向外倾斜
6. 争艳馆是整个花博中最重要的部分之一

7. Night view of snack zone
8. The pergola is folded into several pieces of petals.

7. 零食区夜景
8. 顶棚裂解翻折成数个如花瓣的片状

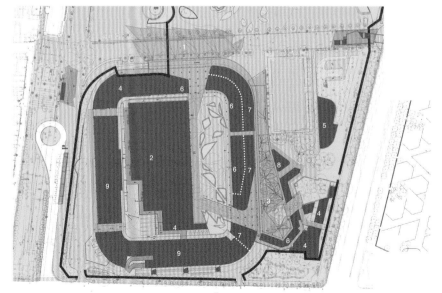

1. Entrance
2. EXPO Dome
3. Snack seating area
4. Restroom
5. Pavilion of New Fashion
6. Public service
7. Shop
8. Dining area
9. Administration office
10. Atrium of EXPO Dome

1. 入口
2. 争艳馆
3. 零食座位区
4. 卫生间
5. 流行馆
6. 公共服务
7. 商店
8. 餐饮
9. 花博办公室
10. 争艳馆中庭

Sun Moon Lake Administration Office of Tourism Bureau

日月潭风景区管理处

Completion date: 2010 **Location:** Taiwan, China **Site area:** 33,340 sqm **Gross floor area:** 6,781.21 sqm **Designer:** Norihiko Dan and Associates **Landscape architect:** Norihiko Dan and Associates (Japan) + Su Mao-Pin architects (Taiwan) **Photographer:** Anew Chen **Client:** Sun Moon Lake Scenic Area Administration

竣工时间：2010年 项目地点：中国，台湾 占地面积：33,340平方米 总楼面面积：6,781.21平方米 设计师：团纪彦建筑事务所 景观建筑师：团纪彦建筑事务所（日本）+苏懋彬建筑师事务所（台湾） 摄影师：安纽·陈 委托人：日月潭风景区管理处

This is one of the projects from an international competition held in Taiwan in 2003 for four representative sightseeing locations in Taiwan called the Landform Series. It is a project for an environment management bureau that houses a visitor centre in the Sun Moon Lake Hsiangshan area.

The site just touches the narrow inlet extending almost south-north at its northern tip, has a narrow opening facing the lake-view direction, and extends relatively deep inland along a road. Looking towards the lake, the lake surface looks like it is cutout in a V shape as mountain slopes close in from both sides. That is, although the site is for the Sun Moon Lake Scenery management bureau, it doesn't have a 180° view of Sun Moon Lake as can be enjoyed from the windows and terraces of the hotels standing on a typically popular site. In most cases with sites like this, the building is positioned on the lake side to secure the greatest view possible, and thus the inland side tends to become a kind of dead space. As the basic policy for the design, the designer's first aim was to propose a new model for a relationship between the building and its natural environment while preserving the surrounding scenery and keeping the inland area from becoming dead space. The second priority was to address the disadvantages of the site whose view of the Sun Moon Lake is not necessarily perfect, and to draw out and amplify the potential advantages.

One way to solve the first problem was to pursue a new relationship between the building and its surrounding landform. Since long ago, buildings have generally been built 'on' landforms, but there have been cases in which they have been built within landforms, such as the early Catholic monasteries of Cappadocia and the Yao Tong settlements along the Yellow River. Due to the fact early modernism negated in totality the methods of self-transformation – including the poche method that belonged to pre-19th century neoclassicism in particular – and demonstrated an inability to adapt to the complex and diverse topography in such areas as east Asia, the designer believe that 20th century architecture actually gave rise to the phenomenon of land development projects that 'flattened' mountains, an approach that is almost synonymous with the destruction of nature. In fact, the very key to linking buildings with landforms lies in these issues that have been ignored by modernism.

In this project, in order to emphasise a sense of horizontality to the architecture, the designer added more soil taken from construction for the foundation to the volume of the building conventionally required, and designed a composition in which the building on the lake side and a sloping mound on the inland side are in gradual and continuous transition. By adopting this composition the continuity is regained between the building and the landform to form an integrated garden rather than having the building sever the landform.

This half-architectural and half-landform project is conceptualised as a stage setting to bring out and amplify a hidden dimension of the scenery and environment of Sun Moon Lake, and at the same time create a new dialogue between the human being and nature that provides another new dimension to this area.

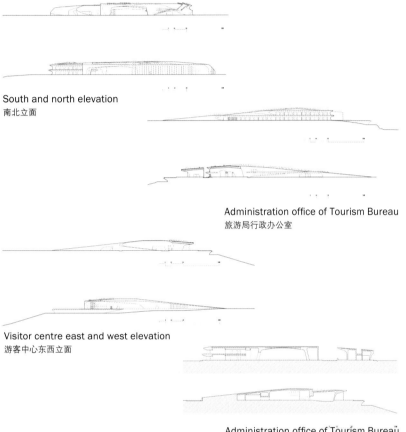

South and north elevation
南北立面

Administration office of Tourism Bureau
旅游局行政办公室

Visitor centre east and west elevation
游客中心东西立面

Administration office of Tourism Bureau
旅游局行政办公室

1. The building is built 'on' the landform
2. The site just touches the narrow inlet extending almost south-north at its northern tip, has a narrow opening facing the lake-view direction, and extends relatively deep inland along a road
3. Skylights on the rooftop

1. 建筑建在景观之"中"
2. 建筑狭窄的入口近乎横跨南北跨度，朝向湖景有一个狭窄的开口，通过一条小路深入内部
3. 建筑顶部的天窗

该项目是台湾2003年在四个代表性观光景点所举办的地形系列竞赛之一,在日月潭香山区打造了一座包含游客中心的环境管理处。

项目位于场地北端,狭窄的入口近乎横跨南北跨度,朝向湖景有一个狭窄的开口,通过一条小路深入内部。两侧的山坡向内闭合,湖面仿佛被切割成V形。尽管场地为日月潭风景处而建,它并没有像酒店一样拥有日月潭的全景窗或露台。在大多数类似的场地,建筑都设在湖畔以保证最好的景色,而内陆一侧则基本被废弃。作为设计的基本原则,设计师的首要目的是打造一个建筑与自然环境关系的新模型,同时保护周边风景,避免内陆一侧成为死角。设计师第二注重的就是正视场地的缺点,认为日月潭的全景并不是必要的,重点要挖掘场地的潜在优势。

第一个问题的解决方案是追求建筑与周边地形之间的全新关系。长久以来,建筑都建在地形之"上",但也有个别案例是建在景观之"中"的,例如:卡帕多西亚早期的天主修道院和黄河流域的窑洞。由于早期现代主义全盘否定了自我改造的方法——特别是19世纪前新古典主义的壁龛,并且论证了东亚这样的地势无法适应景观与建筑融为一体,设计师认为20世纪的建筑促生了许多"铲平"山头的土地开发项目,对自然进行了破坏。事实上,现代主义完全忽略了建筑与地形之间的最重要的联系。

在本项目中,为了突出建筑的水平感,设计师从地基建造中挖掘出更多的土壤用于建筑结构建造,在湖畔一侧设计了一座建筑,在内部一侧则设计了一块坡地,二者之间形成了流畅的过渡。这种组合让建筑与地势之间重新获得了连续性,形成了一个综合的花园,而不是让建筑与景观分离。

这座半建筑、半景观项目为日月潭的风景和环境带来了隐藏维度,同时也在人类和自然之间建立了新的对话,为整个区域提供了新的生机。

4. The building on the lake side and a sloping mound on the inland side are in gradual and continuous transition.
5. The architecture is on the landform, inseparable from the surroundings.
6. Details of the building
7. Building in construction

4. 湖畔一侧是建筑，在内部一侧则是坡地，二者之间形成了流畅的过渡
5. 建筑在景观之中，不与周围景色分离
6. 建筑细部
7. 建设中的建筑

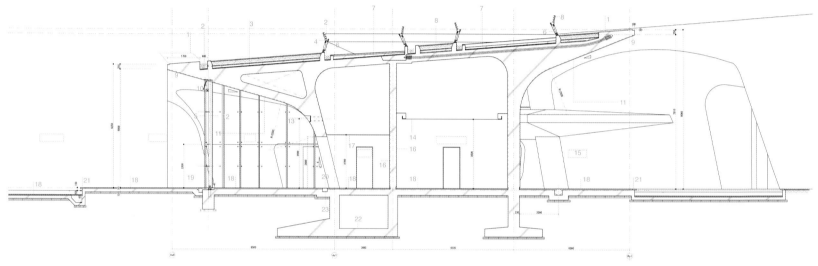

1. Fast Curing Polyurethane Waterproof Mortar, Painted Milky White Colour
2. Crushed Stones Waterproof Mortar Fast Curing Polyurethane Waterproofing
3. Grass Soil t=200 Drainage Material t=30 Retaining Wall t=100 Insulation t=30 Fast Curing Polyurethane Waterproofing Concrete Surfacing
4. HL Stainless Steel Handrail
5. Reinforced Glass t=15 (DPG joint method)
6. Stainless Frame within Stainless Net
7. Tiles t=10 Mortar Retaining Wall t=100 Insulation t=30 Fast Curing Polyurethane Waterproofing Concrete Surfacing
8. PC Strand Cable
9. Water-cutting Joint
10. Smoke Ventilation Window: Stainless Steel Sash Powder Coating Paint
11. Wall and Ceiling: Exposed Concrete in Taiwan Cedar Wood Pattern Water Repellent Paint, Milky White Colour
12. Mullion: Steel Plate Powder Coating Paint
13. Air Outlet 150 @1800
14. Plasterboard t=12.5*2 Putty Repair, White Paint
15. Corridor
16. RC, Mortar Repair, White Paint
17. Calcium Silicate Board t=6VP
18. Tiles t=10 Mortar t=40
19. Stainless Gutter Cover
20. Tempered Glass t=10 Diffusion Glass
21. Stone Border
22. Waterproof Mortar
23. Built-in Lighting

1. 快速固化聚氨酯防水灰浆，乳白色漆
2. 碎石防水灰浆，快速固化聚氨酯防水层
3. 玻璃土 t=200，排水材料 t=30，挡土墙 t=100，保温层 t=30，快速固化聚氨酯防水混凝土面
4. HL不锈钢扶手
5. 加固玻璃 t=15（DPG接合法）
6. 不锈钢网内的不锈钢框
7. 瓷砖 t=10，砂浆挡土墙 t=100，保温层 t+30，快速固化聚氨酯防水混凝土面
8. PC绞合电缆
9. 水切接合
10. 通风窗：不锈钢框，粉末涂料
11. 墙壁和天花板：露石混凝土，采用台湾雪松木纹防水涂料，乳白色
12. 竖框：钢板，粉末涂料
13. 排气口150 @1800
14. 石膏板t=12.5*2，油灰修补，白色涂料
15. 走廊
16. 钢筋混凝土，灰泥修补，白色涂料
17. 硅酸钙板 t=6VP
18. 瓷砖 t=10，砂浆 t=40
19. 不锈钢排水盖
20. 钢化玻璃t=10，漫射玻璃
21. 石线
22. 防水灰浆
23. 嵌入式照明

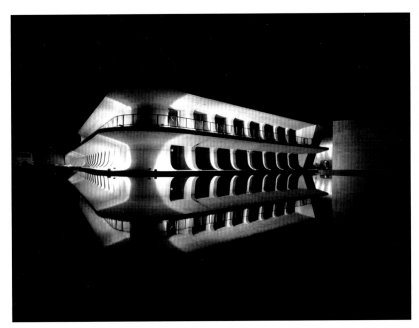
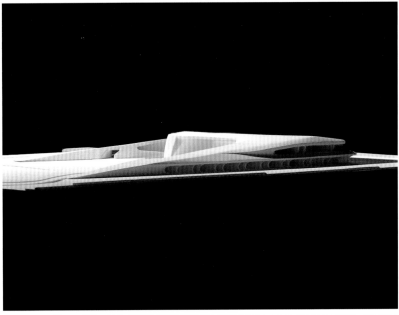
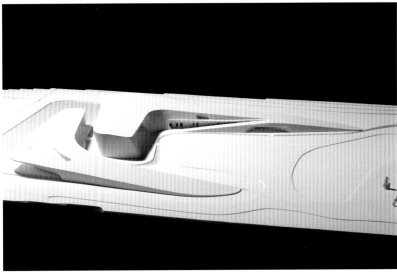

1. Information
2. Café
3. Exhibition room 1
4. Multimedia briefing room
5. Exhibition room 2
6. Exhibition room 3
7. Office entrance Hall
8. Office
9. Conference room
10. Main conference room
11. Staff dining room
12. Director's room
13. Guest room
14. Vice Director's room

1. 信息区
2. 咖啡厅
3. 展览厅1
4. 多媒体接待室
5. 展览厅2
6. 展览厅3
7. 办公室前厅
8. 办公室
9. 会议室
10. 主会议室
11. 员工餐厅
12. 总监室
13. 客房
14. 副总监室

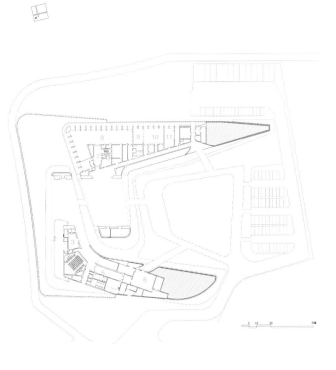
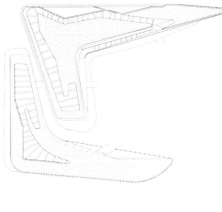
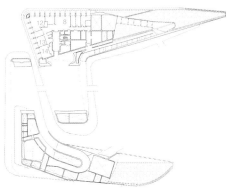

Edithvale Seaford Wetlands Discovery Centre

艾迪斯维尔西弗德湿地发现中心

Completion time: 2012 **Location:** Victoria, Australia **Designer:** Minifie van Schaik Architects
Photographer: Peter Bennetts

竣工时间：2012年 项目地点：澳大利亚，维多利亚 设计师：MvS建筑事务所 摄影师：彼得·波奈斯

The Edithvale Seaford Wetlands Discovery Centre, for Melbourne Water, is perched overlooking the Edithvale wetlands, a remnant of the once extensive and ecologically rich Carum Carrum swamp – now recognised by a RAMSAR listing for being of international significance.

The centre provides an interpretive experience for visitors to illuminate the vital and complex history and workings of this urban wetland in and its role in the water-cycle.

It stands pelican-like above the wetlands rendering it safe from flood waters. A long ramping approach allows the visitor to journey upwards for a view of the flora and fauna beyond before leading to an entry airlock comprised of a sculpted internal water storage unit.

The interior contains, offices, toilets and an interpretive gallery offering views of the wetlands through panoramic windows whose glazing has been designed to minimise vision into the building, and raked to ensure that external reflections are always of the ground, never the sky, lest birds terminally confuse reflection for reality.

Windows in the floor of the display centre allow views of the phragmites and diverse wildlife within the fenced wetlands boundary which stretches beneath the conical legs of the building.

Solar panels, heat-pumps, floor-grate supply and passive extraction, double glazing, motion controlled lighting, high levels of insulation and thermal mass together allow building operations to approach carbon neutral. Combined with a composting sewerage system and an internal water storage allowing the centre to attract an equivalent of six Green Stars.

Ironically, a clause in the green star rating system excludes anything built within a given radius of a RAMSAR wetlands from receiving a green star rating. So, while the centre is designed to improve general knowledge and attitude to this wetland, and by inference wetlands in general, the Green Star clause designed to protect wetlands precludes the rating system from recognising the centre's significant sustainable design features. The centre is expressive of its location in a hybrid urban and wetland condition and has been designed to embody an approach to the relationship between humans and nature that includes both, rather than idealising one over the other.

The bright orange of the undercroft is designed to be reminiscent of the birds the centre will educate the public about. The same birds around whose migratory patterns the construction period was delicately tailored. In certain lighting conditions the colour of the undercroft reflects earthwards and the building emits an encouraging glow as if attempting to levitate.

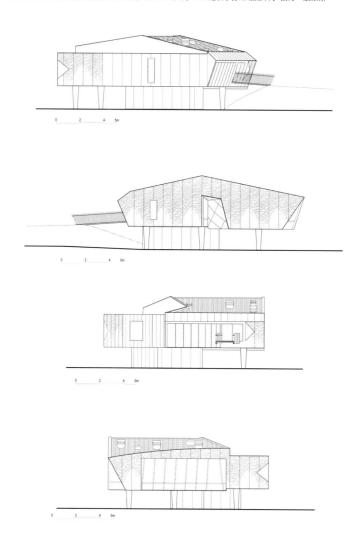

1. The centre is expressive of its location in a hybrid urban and wetland condition.
2. The panoramic window is slightly raked.
3. Entryway and fenced wetlands boundary

1. 建筑位于城市和湿地的混合环境中
2. 展廊的全景窗略微倾斜
3. 建筑入口通道和湿地围栏边界

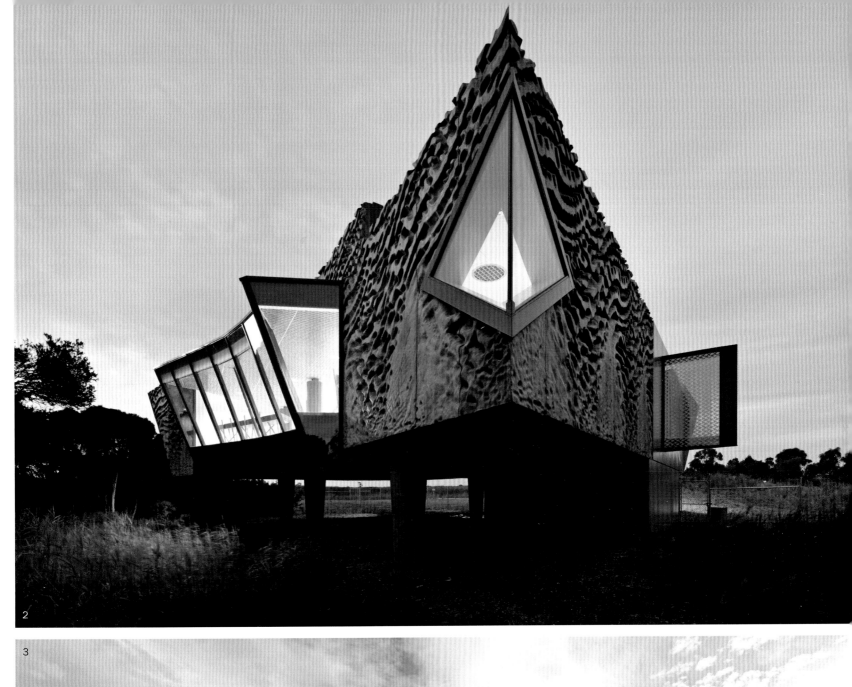

艾迪斯维尔西弗德湿地发现中心俯瞰着艾迪斯维尔湿地——广阔而丰富的卡拉姆沼泽的残余，现在已经被国际湿地公约秘书处列为国际湿地名录。

中心为游客提供了丰富的讲解体验，展示着这片城市湿地鲜活而负责的历史和作品，以及它在水循环中的重要性。

建筑像鹈鹕一样矗立于湿地之上，以免受洪水侵袭。长长的坡道让游客能够远眺远方的动植物，引导着他们进入包含内部水仓库的入口气闸。

建筑内部设有办公室、洗手间和解说展廊。展廊通过全景窗向游客展示湿地的景色，其设计略微倾斜，以保证窗口只展示地面的景象，而不是天空，以免鸟类混淆了现实的倒影。

展示中心地面的窗户让人们可以观察芦苇和湿地围栏内各种各样的野生物，围栏的边界在建筑下方的圆柱腿下方。

太阳能板、热泵、地热供暖和被动抽气、双层玻璃、运动控制照明、高度隔热和保温层共同使得建筑运营实现了碳平衡。发现中心与堆肥污水系统和内部水库相连，使它达到了绿色之星六星级标准。

由于绿色之星评价系统的一条条款拒绝为任何建在国际湿地公约名录湿地内的建筑评级，尽管中心的设计提升了公众常识，对湿地做出了贡献，绿色之星仍然没有为中心颁发可持续设计认证。发现中心位于城市和湿地的混合环境中，体现了对人类与自然之间关系的深入思考，使二者处于平等的地位。

亮橘色的圆顶地下室让人想起了发现中心附近的鸟类。迁徙方式与施工期相同的鸟类受到了精心照料。在特定的灯光条件下，地下室的色彩能够反射地面，建筑仿佛浮在空中一样。

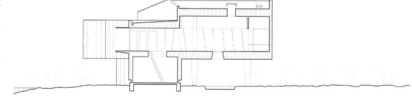

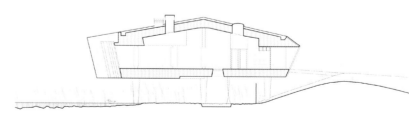

4.5. The conical legs prevent the building from flood.
6. Night view of the entrance

4、5.建筑下方的圆柱腿使其免受洪水侵袭
6. 建筑入口夜景

7. Exhibition centre
8. The windows are raked to ensure that external reflections are always of the ground.
9. Interior restroom

7. 展示中心
8. 倾斜的窗口只展示地面的景象
9. 室内洗手间

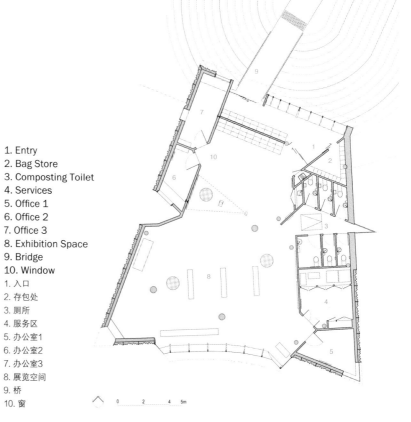

1. Entry
2. Bag Store
3. Composting Toilet
4. Services
5. Office 1
6. Office 2
7. Office 3
8. Exhibition Space
9. Bridge
10. Window

1. 入口
2. 存包处
3. 厕所
4. 服务区
5. 办公室1
6. 办公室2
7. 办公室3
8. 展览空间
9. 桥
10. 窗

8

9

大沙湾海滨浴场 Dashawan Beach Facility

Completion date: 2009 **Location:** Liandao Island, Lianyungang, China **Site area:** 20,758 sqm **Building area:** 7,761 sqm **Designer:** Scenic Architecture Ofce **Design Team:** Zhu Xiaofeng, Cai Jiangsi, Xu Lei, Xu Ye, Ding Xufen **Structural & MEP:** Shanghai Aco International Design Consultant Co. **Photographer:** Shen Zhonghai **Client:** Liandao Seafront Resort Committee **Programme:** Locker & Shower, Restaurant & bar, Entertainment, Hotel

竣工时间：2009年 项目地点：中国，连云港市，连岛 占地面积：20,758 平方米 建筑面积：7,761平方米 设计师：山水秀建筑事务所 设计团队：祝晓峰、蔡江思、许磊、许曳、丁旭芬 结构与机电设计：上海原构国际设计咨询公司 摄影师：沈忠海 业主：连岛海滨度假区管委会 建筑功能：更衣、餐饮、娱乐、酒店客栈

The site locates on the east seafront of Liandao island of Lianyungang, an emerging harbour city booming in the middle section of the coastline of China. The beach hosts about 20,000 swimmers everyday in the peak time of summer. The design team was commissioned to design new facilities for bath service, restaurant & bars, fitness, entertainment and accommodations to meet the needs of increasing crowd.

Facing east to the Pacific Ocean, the building lay on the hillside with 3 Y-shape slabs. The stacking and setback platforms of the slabs distribute pedestrian flows to different levels from the main entry of the beach, while offering terrific views for all floors of the building. The Y-shape provides opportunities for different functional volumes to interact with each other and creates dynamic experience of open spaces on and below the roofs. These hundreds-meter-long slab buildings establish a scale dialogue with both the coastal mountain and wave topography of the ocean, hence to present an ambition of the city to develop and grow.

连云港是中国海岸线中部正在崛起的一座海港城市，基地位于连云港北部连岛度假区的东海岸。这片"江苏省最好的沙滩"在夏季高峰时每天吸引2万名游客。我们的设计为不断增加的客流量提供了新的更衣设施、餐饮、酒廊、健身中心、娱乐及住宿场所。

这座建筑朝东面向太平洋，由3块搁置在沙滩后方山坡上的"Y"形板体组成。板体之间的叠层和退台将来自南侧入口的人流引导到不同的平台，并为所有楼层提供了壮丽的海景视线。"Y"形板体的上下斜坡形成了多样化的户外开放空间，为不同的功能活动提供了交流机会。这些超过百米长度的板状建筑与自然界的山坡和海浪同时建立了尺度上的关系，并表达了这座城市崛起的雄心。

Cross section
横剖面

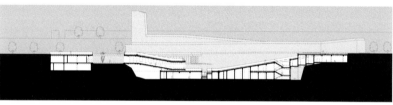

Longitudinal section
纵剖面

1. Sea to the east
2. The pavilion is composed with three "Y"s
3. Layered roof deck

1. 东向海景
2. 由3个Y形建筑组成
3. 叠层的屋顶平台

1

4. Beach to the northeast
5. Beach to the southeast
6. Roof of the guest house
7. Interior design with panoramic windows

4. 东北向海滩
5. 东南向海滩
6. 客栈屋顶
7. 全景玻璃窗的室内设计

Middle floor plan
中部平面

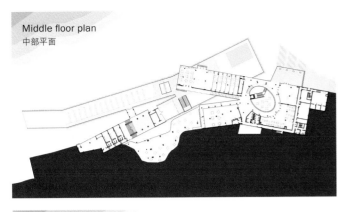

Lower floor plan
下部平面

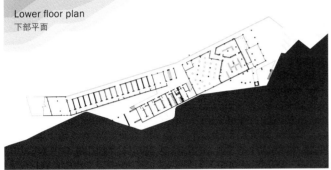

Upper floor plan
上部平面

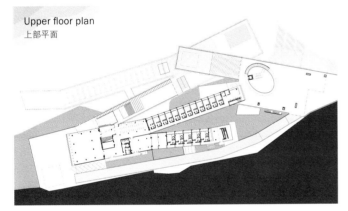

发现绿色公园景观建筑 Architecture of Discovery Green

Completion date: 2008 **Location:** Houston, Texas, USA **Site area:** 10.12 ha **Designer:** PageSoutherlandPage, Hargreaves Associates, Lauren Grifth Associates **Photographers:** Eric Laignel, Chris Cooper, John Gollings

竣工时间：2008年 项目地点：美国，休斯顿 占地面积：10.12公顷 设计师：PSP建筑公司、哈尔格里维斯事务所、劳伦·格里菲斯事务所 摄影师：埃里克·莱格尼尔、克里斯·库珀、约翰·高林斯

Located in an area of downtown Houston once dominated by large public facilities (Toyota Centre, Minute Maid Park and the George R. Brown Convention Centre) and their often-empty parking lots, Discovery Green is the result of a partnership between the City of Houston and private philanthropists.

The Gold LEED certified, 12-acre park includes two restaurants, a park administration building, underground parking for more than 600 vehicles and numerous site features. The three primary buildings on the site – the Lake House café, the park building and the Grove Restaurant – parallel the grove of existing live oaks and reinforce their linear character. Each building is composed of long, thin volumes that draw activity from the major north/south promenade deep into the park on either side.

It was extremely important that Discovery Green be specifically of Houston, and constructed with identity-defining and sustainable local materials. In addition to using native and regionally appropriate plant materials throughout the park, a distinctive red-orange Gulf Coast brick is employed in a strongly horizontal coursing pattern to reflect the emphatic flatness of the clay geology of the region. The design focused on making landscape-oriented buildings that would blend seamlessly with the outdoor environment and would be respectful of natural forces and phenomena. There is as much outdoor space in the buildings as indoor space. Park buildings are characterised by expansive glass faces on the north exposure, capturing natural lighting and creating contiguous indoor/outdoor relationships, while large shaded outdoor verandas on southern exposures reduce solar heat gain and encourage outdoor seating and gathering by providing shelter from Houston's characteristic hot sun and periodic downpours.

The Grove Restaurant is dominated by a long, thin dining room that nestles under the boughs of the Oak Allee. Tall glass walls toward the trees and at each end open the room generously to the park, while a richly textured brick volume housing kitchen and service functions anchors the room on the street side. The upper level of the restaurant is predominantly a shaded outdoor dining terrace accessed by broad staircases at the east and west ends.

The administration building and the Lake House café have deep, shady porches that dominate their south faces. Carefully designed to create a shield from hot south and west sun, the porch roofs pitch up to the north to achieve balanced daylight for the outdoor spaces below as well as to induce air movement, drawing warm air up and out. The south-facing roofs of the administration building and the Lake House porches support 256 photovoltaic collectors that provide 8% of the power needed for the park.

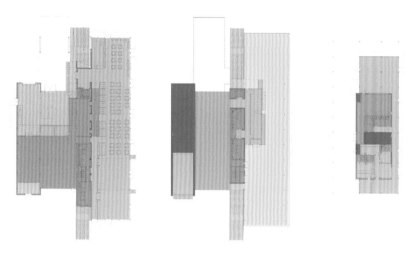

Grove and Lake House Plans
树林餐厅和湖屋餐厅平面图

1. View across the lake
2. Picnic lawn
3. Night view of the administration building and the lake house restaurant

1. 从湖对面看建筑
2. 野餐草坪
3. 行政楼和湖屋餐厅夜景

4. Ariel view of the park in night
5. Carefully designed to create a shield from hot south and west sun, the porch roofs pitch up to the north to achieve balanced daylight for the outdoor spaces below as well as to induce air movement, drawing warm air up and out.
6. The vehicular entry for the below-grade garage is nestled into a land berm that reflects the shape of the garage ramp as it descends into the ground, creating an amphitheater and play area above.

4. 公园夜间鸟瞰图
5. 门廊屋顶向北侧拉高，为下方的露天空间带来了均衡的阳光，同时也促进了空气的流通，让热气上升、流出
6. 地下车库的入口嵌入了坡台之中，坡台反映出车库坡道的造型，形成了露天剧院及其上方的游乐区

发现绿色公园位于休斯顿市中心，场地曾经被许多大型公共设施（如丰田中心、美之源公演和乔治·R·布朗会议中心）和它们空置的停车场占据。发现绿色公园是休斯顿市政府和私人慈善家共同合作的成果。

这座获得绿色建筑金奖认证的公园包含两座餐厅、一座公园行政楼、600多个车位地下停车场和大量场地景观设施。场地上的三座主要建筑——湖屋咖啡厅、停车楼和树林餐厅与橡树林平行，凸显了它们的直线特色。每座建筑都有细长的结构组成，从公园南北两侧的散步长廊吸引人前来进行活动。

最重要的是，发现绿色公园对休斯顿城有着特殊的意义，它由特色鲜明的可持续材料建造而成。除了利用本地和本地区的植物材料之外，公园内独特的红橙色海湾呈水平方向排列，突出了当地黏土地质的平坦。设计聚焦于打造以景观为中心的建筑，使其与户外环境紧密地结合起来，同时尊重自然力量和现象。建筑的室外空间与室内空间几乎相等。停车楼北侧拥有独特的玻璃面，捕捉了自然光并打造了连续的室内外体验。而南侧的露天遮阳游廊则减少了太阳热增量，鼓励人们在户外休息和集会，使他们远离休斯顿独有的烈日和暴雨。

树林餐厅拥有一个狭长的就餐区，在橡树大道的树枝下若隐若现。高大的玻璃墙让餐厅面向公园开放，而面向街道的纹理丰富的砖结构内则设有厨房和服务区。酒店的二楼是一个露天就餐平台，可以由东西两侧宽大的楼梯进入。

行政楼和湖屋咖啡厅南面拥有深邃而阴凉的门廊。门廊的设计精心打造了一个远离烈日和夕阳的空间，门廊屋顶向北侧拉高，为下方的露天空间带来了均衡的阳光，同时也促进了空气的流通，让热气上升、流出。行政楼和湖屋门廊朝南的屋顶上设有256个光电伏收集器，为公园提供了8%所需的电力。

Awards:
2009 AIA Austin Design Award
2010 Associated Masonry Contractors of Houston Golden Trowel Excellence Award – Industrial/Commercial
2010 Green GOOD DESIGN Award
2008 USGBC-Houston Outstanding Environmental Project

获奖情况：
2009美国建筑师协会奥斯汀设计奖
2010休斯顿砖承包商金泥刀奖——工业/商业类
2010绿色好设计奖
2008美国绿色建筑委员会休斯顿优秀环境项目

7. West façade of the Grove on a sunny fall afternoon
8. The glass walls on the north façade of the Grove open generously to the park and the mature oak trees which were saved during the construction of the park
9. View across the Great Lawn toward the park building and Lake House. Pedestrian access to the below-grade parking garage is hidden inside one of Margo Sawyer's colourful aluminium boxes

7. 阳光明媚的午后，树林餐厅的西立面
8. 树林餐厅北侧的玻璃墙朝向公园和高大的橡树；橡树在公园建设中得以保留下来
9. 跨过大草坪看停车楼和湖屋咖啡厅；地下车库的行人入口隐藏在彩色的铝盒结构之中

1. Entry plaza	1. 入口广场
2. Picnic lawn	2. 野餐草坪
3. Veranda	3. 门廊
4. Playground	4. 运动场
5. Garden	5. 花园
6. Administration building	6. 行政楼
7. The lake house restaurant	7. 湖屋餐厅
8. Terrace	8. 平台
9. Gateway fountain	9. 入口喷泉
10. Model boat basin	10. 模型船水池
11. Stage	11. 舞台
12. Performance space	12. 表演空间
13. Great lawn	13. 大草坪
14. Garage stairs	14. 车库楼梯
15. Kinder lake	15. 辛德尔湖
16. Waterside landing & garden	16. 水边码头和花园
17. Jogging trail	17. 缓跑道
18. Garage elevator	18. 车库电梯
19. Garage entry	19. 车库入口
20. The Brown foundation promenade	20. 布朗喷泉长廊
21. The grove restaurant	21. 树林餐厅
22. Event lawn	22. 活动草坪
23. Fountain	23. 喷泉
24. Pine grove	24. 松树林
25. Plaza	25. 广场
26. Andrea and Bill White promenade	26. 安德里亚和比尔·怀特长廊

10. Entrance of the Grove Restaurant
11. View from corner of the Lake House toward administration building
12. The 'Treehouse' has shaded decks that nestle into the upper branches of the live oaks. In inclement weather, these rooms can be fully enclosed by sliding glass panels

10. 树林餐厅的入口
11. 从湖屋餐厅的一角看行政楼
12. "树屋"的遮阳平台位于橡树的槲树林的上层枝干之上；恶劣天气时，房间的滑动玻璃板可以完全封闭

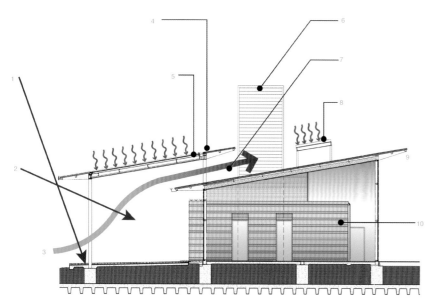

1. Summer sun excluded
2. Winter sun penetration
3. Cross-ventilation warm air rises and induces flow
4. Shading structure
5. Optimum photovoltaic panel orientation
6. Ventilation chimney
7. Balance natural lighting to reduce glare
8. Optimum solar water heating panel origination
9. Diffuse north light for delighting
10. High thermal mass for temperature stability

1. 遮挡夏日阳光
2. 冬日阳光渗入
3. 交叉通风，暖空气上升，引导空气流动
4. 遮阳结构
5. 最佳光电伏板朝向
6. 通风烟囱
7. 平衡自然光，以减少刺眼的光线
8. 最佳太阳能热水器朝向
9. 从北面散射而来的光线令人心情愉悦
10. 建筑内核的高质量蓄热体能保持温度稳定性

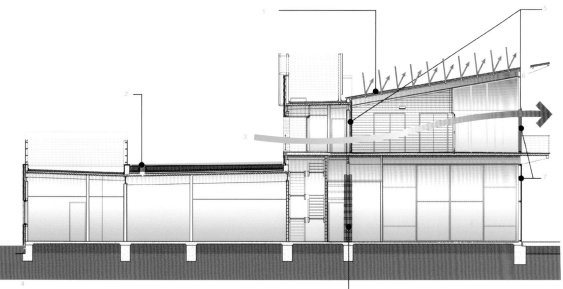

1. Reflective metal roof
2. Green roof
3. Cross-ventilation warm air rises and induces flow
4. Masonry construction at south portion of the building to protect from intense solar radiation
5. Operable glass wall
6. Outdoor room
7. Diffuse north light for delighting
8. High thermal mass in building core for temperature stability

1. 反光金属屋顶
2. 绿色屋顶
3. 交叉通风,暖空气上升,引导空气流动
4. 建筑南部的砖石结构保护其不受强烈的太阳辐射
5. 可控式玻璃墙
6. 露天区域
7. 从北面散射而来的光线令人心情愉悦
8. 建筑内核的高质量蓄热体能保持温度稳定性

图书在版编目（CIP）数据

风景建筑 /（卢森堡）瓦格纳，（卢森堡）韦伯编；常文心译. -- 沈阳：辽宁科学技术出版社，2013.5
ISBN 978-7-5381-8006-0

Ⅰ. ①风… Ⅱ. ①瓦… ②韦… ③常… Ⅲ. ①园林建筑－建筑设计 Ⅳ. ①TU986.2

中国版本图书馆CIP数据核字(2013)第068441号

出版发行：辽宁科学技术出版社
　　　　　（地址：沈阳市和平区十一纬路29号　邮编：110003）
印　刷　者：利丰雅高印刷（深圳）有限公司
经　销　者：各地新华书店
幅面尺寸：245mm×290mm
印　　张：34
插　　页：4
字　　数：90千字
印　　数：1～1200
出版时间：2013年 5 月第 1 版
印刷时间：2013年 5 月第 1 次印刷
责任编辑：陈慈良　吴　杨
封面设计：周　洁
版式设计：周　洁
责任校对：周　文
书　　号：ISBN 978-7-5381-8006-0
定　　价：298.00元

联系电话：024-23284360
邮购热线：024-23284502
E-mail: lnkjc@126.com
http://www.lnkj.com.cn
本书网址：www.lnkj.cn/uri.sh/8006